PLANT CONTAMINATION

Modeling and Simulation
of Organic Chemical Processes

Edited by

Stefan Trapp and J. Craig Mc Farlane

LEWIS PUBLISHERS

Boca Raton   Ann Arbor   London   Tokyo

Library of Congress Cataloging-in-Publication Data

Plant contamination : modeling and simulation of organic chemical
processes / edited by Stefan Trapp, Craig Mc Farlane.
    p.   cm.
Includes bibliographical references and index.
ISBN 0-56670-078-7
    1. Plants, Effect of xenobiotics on--Mathematical models.
I. Trapp, Stefan.   II. Mc Farlane, Craig, 1936-
QK753.X45P58   1995
581.2—dc20

94-9734
CIP

Preface

This book presents a description of the physiological and anatomical principles, and the chemical and physical factors that determine uptake, translocation, accumulation, loss, and metabolism of anthropogenic chemicals in plants. Mathematical models are presented that integrate this diverse knowledge into realistic and useful information. The purpose of the book is to provide the reader with recently developed methods of estimating the behavior of xenobiotics in the plant-air-soil system. This information is essential in the hazard assessment of new and existing chemicals. Besides the ecotoxicological aspects, some economic advantages can also be drawn: the efficacy of pesticides is strongly influenced by their transport to sites of action. Reduction of chemical contamination in food webs can also be evaluated by using information in this book.

Classically, plant physiologists have been concerned with plant growth and development and thus have restricted their studies to uptake, synthesis, and transport of water, mineral nutrients, sugars, and certain organic biochemicals associated with hormonal control or biomass synthesis. With the advent of chemical agriculture, there has been great attention to plant interactions with herbicides and pesticides. However, most of this work has dealt with efficacy and residues in harvestable tissues and has ignored the basic principles that determine uptake, persistence, and fate. This book is intended to provide a discussion which outlines current progress in basic concepts and the synthesis of various areas of knowledge into simulation models. Expert authors in the fields of biology, chemical engineering, ecology, and physics give this book an interdisciplinary view of an important topic.

The book is directed to all persons interested in plant contamination. Since it covers concepts of physics, chemistry, mathematics, ecology, physiology, and biophysics, it will be of interest to readers from many disciplines. It will be especially useful to persons engaged in regulation, use, dissemination, and manufacture of toxic chemicals. Models useful in predicting plant contamination are described.

About the Editors

Stefan Trapp, Ph.D., studied Geoecology at the University of Bayreuth, Germany. While at the university his primary interest was focused on environmental modeling. He worked at the National Center for Environmental Protection (GSF) in Munich from 1987 to 1992.

While his first interest was in aquatic systems, he soon recognized that the knowledge of terrestrial systems was much smaller. As a result, he developed, in his Ph.D. thesis, a simulation model for the uptake of organic chemicals into plant. In 1992 he became affiliated with the University of Osnabrück where he has been teaching "Applied System Science".

Dr. Trapp has published more than 30 papers, most of them on the subject of the development, validation, and application of environmental fate models. He is a member of the German Society for Geoecology. Since 1994 he has been a member of the Institute of Environmental Systems Research where he continues his work in this area.

J. Craig Mc Farlane, Ph.D., is an environmental scientist at the Environmental Research Laboratory, Corvallis, Oregon. He received a Ph.D. in Plant Physiology from the University of California at Riverside in the Vegetable Crops Department. Throughout his career with the U.S. Environmental Protection Agency, he has conducted research on the contamination of plants. His most recent work has centered on the patterns and mechanisms of root uptake, translocation, accumulation, and degradation of organic chemicals, using whole plants in controlled environments.

He served on science advisory boards for NASA and EPA, published over 80 scientific articles, and has delivered guest lectures throughout the world on aspects of plant contamination. Dr. Mc Farlane has served on the editorial boards of numerous scientific journals, chaired the Committee for Controlled Environments and Growth Chambers for the American Society for Horticultural Science, and has organized various international seminars and symposia.

During a sabbatical study leave with Prof. Dr. Michael Matthies at the GSF in Munich, Germany, he met and worked with Dr. Stefan Trapp. They became interested in the idea of mechanistic models useful in predicting plant contamination. Thus, this book was conceived and born.

Contributors

Herwart Behrendt
Projektgruppe
Umweltgefährdungspotentiale von
  Chemikalien
GSF-Forschungszentrum für
  Umwelt und Gesundheit
D-85758 Oberschleißheim, Germany

Richard H. Bromilow
Department of Biological and
  Ecological Chemistry
Institute of Arable Crops Research
Rothamsted Experimental Station
Harpenden, Herts AL5 2JQ
United Kingdom

Keith Chamberlain
Department of Biological and
  Ecological Chemistry
Institute of Arable Crops Research
Rothamsted Experimental Station
Harpenden, Herts AL5 2JQ
United Kingdom

Dieter Komoßa
GSF-Forschungszentrum für
  Umwelt und Gesundheit
Institut für Biochemische
  Pflanzenpathologie
Ingolstädter Landstraße 1
D-85764 Oberschleißheim, Germany

Christian Langebartels
GSF-Forschungszentrum für
  Umwelt und Gesundheit
Institut für Biochemische
  Pflanzenpathologie
Ingolstädter Landstraße 1
D-85764 Oberschleißheim, Germany

J. Craig Mc Farlane
U.S. Environmental Protection Agency
Environmental Research Laboratory
200 S.W. 35th Street
Corvallis, Oregon 97333

Donald Mackay
Institute for Environmental Studies
University of Toronto
Toronto, Ontario, Canada

Michael Matthies
Institute of Environmental Systems
  Research
University of Osnabrück
D-49069 Osnabrück, Germany

Sally Paterson
Institute for Environmental Studies
University of Toronto
Toronto, Ontario, Canada

Markus Riederer
Julius-von-Sachs-Institut für
  Biowissenschaften
Lehrstuhl für Botanik II
Universtät Würzburg
Mittlerer Dallenbergweg 64
D-97082 Würzburg, Germany

Heinrich Sandermann, Jr.
GSF-Forschungszentrum für
  Umwelt und Gesundheit
Institut für Biochemische
  Pflanzenpathologie
Ingolstädter Landstraße 1
D-85764 Oberschleißheim, Germany

Stefan Trapp
Institute of Environmental Systems
  Research
University of Osnabrück
D-49069 Osnabrück, Germany

Contents

CHAPTER **1**

Introduction

Stefan Trapp and J. Craig Mc Farlane

TABLE OF CONTENTS

For most of human history, pollution was primarily human waste along with litter and smoke, but with the advent of modern chemistry, the scene changed dramatically. Rachel Carson (1962)[1] was one of the first to draw attention to the serious consequences of our carelessness. Environmental monitoring has since revealed many uncalming results, including the identification of a list of persistent chemicals which have been found over the entire globe, not only in industrial areas, but in our food, our backyards, remote wilderness areas, and even in arctic environments. Anthropogenic chemicals have been found in water, sediments, fish, plants, birds, and in humans.

Most people don't regard residual concentrations of some pesticides as threatening; rather we rely on government regulations to protect us from harmful levels of toxic or carcinogenic chemicals. However, when faced with a real or imagined threat, consumers panic and suppliers remove various vegetable products from public markets as was demonstrated by the publicity about Alar®* (a plant growth regulator) in/on apples in the U.S. in 1989. Most governments attempt to regulate the production, distribution, and use of chemicals in a manner that protects health and the environment at the same time as

*Uniroyal Chemical, Fresno, CA.

1-56670-078-7/95/$0.00+$.50
© 1995 by CRC Press, Inc.

protecting agricultural production and the profits of chemical companies. This balancing act is based on consideration of measured productivity and profit against assumed and estimated measures of safety. Exposure is estimated by environmental distribution models and effects on the basis of (representative?) toxicity tests.

Pesticides are designed to protect plants from competition (herbicides) or from predators (insecticides, nematocides, and fungicides). Since these chemicals are intended to be applied directly to plants or to soils, plant contamination is an important concern of chemical companies and is considered when the chemical is registered or licensed for use. Various industrial chemicals are not intended for such application and become contaminants only in response to accidents, misuse, or inadvertent dispersal such as volatilization, non-degradability in waste treatment, or other unanticipated phenomena.

Studies to learn the patterns of chemical release, persistence, environmental fate, and the impact on ecological systems and human health have proliferated dramatically. In an effort to summarize this literature, a search was done in cooperation with the German Advisory Committee on Existing Chemicals of Environmental Relevance[2] to identify chemicals that constitute a potential for food contamination. A total of 68 chemicals was selected. Among these were 21 polycyclic aromatic hydrocarbons (PAHs), 15 halogenated aromatics, 11 other substituted aromatics, phenols, biphenyls, aliphatics, and more. The information summarized in Figures 1 to 8 came from 551 printed documents (books, articles, reports) and 265 computer databases which yielded a total of 7698 statements about concentrations of 68 anthropogenic chemicals in the environment.

Concentrations are given for plants (corn, fruit, and vegetable), fish (including crustacea), meat, urban soil, sediment, drinking water, surface water (mostly rivers or lakes), urban air, and rural air. All values were recalculated to represent nanograms per liter (water), nanograms per cubic meter (air), and nanograms per kilogram (sediments and food). This allows a direct comparison of the occurrence of the chemicals in these media. The concentrations are expressed on a log scale (base 10) because of the large range. The distribution of results is mostly lognormal; thus, the geometric mean is given instead of the arithmetic mean (rue symbol). Adjacent to the y-axis, the number of hits (values about concentrations that were found) is given within brackets.

I. ANTHRACENE AND PHENANTHRENE

Anthracene is a three-ring PAH. Like most of the PAHs it is found in tar and as a product of incomplete combustion. From Figure 1, note that the highest concentrations were found in sediments (up to 70 mg/kg) and urban soil (only one value, 1 mg/kg). Concentrations in water and air samples (10 ng/l and 1 ng/m^3, respectively) were typically much lower than in samples from other compartments. Phenanthrene has a similar distribution pattern (Figure 2).

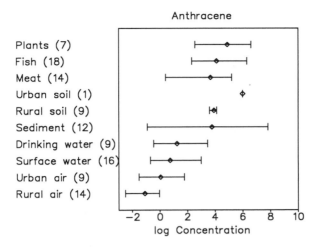

FIGURE 1. Concentration ranges of anthracene in the environment.

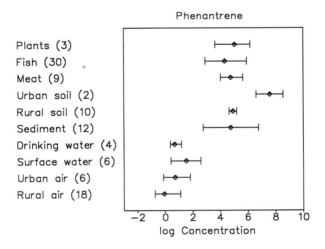

FIGURE 2. Concentration ranges of phenanthrene in the environment.

II. OTHER POLYCYCLIC AROMATIC HYDROCARBONS

Benzo(a)pyrene has received a great deal of attention because of its carcinogenicity. Fluoroanthene and pyrene were also represented by enough data to complete our chart. Concentrations in various compartments vary dramatically, but in general the concentrations in food, soils, and sediments are higher than in water (Figures 3 to 5).

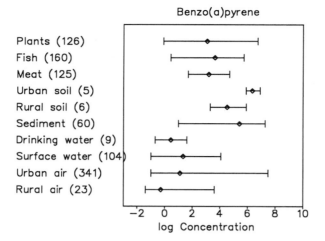

FIGURE 3.   Concentration ranges of benzo(a)pyrene.

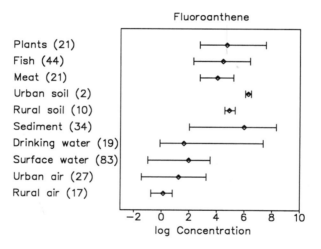

FIGURE 4.   Concentration ranges of fluoroanthene.

III. CHLOROORGANIC COMPOUNDS

This group of high importance was not represented by sufficient data to complete the charts. Incomplete data sets are nevertheless shown for di-, tri-, and hexachlorobenzene (Figures 6 to 8). Notably lacking was information on plant, meat, and soil concentrations. Hexachlorobenzene (HCB) was represented by values for all media except air.

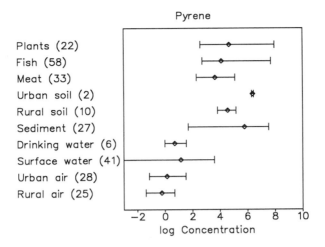

FIGURE 5. Concentration ranges of pyrene.

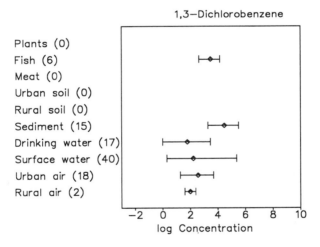

FIGURE 6. Concentration ranges of 1,3-dichlorobenzene.

IV. WHAT CAN BE LEARNED FROM THIS COMPARISON?

First, it is obvious that most measurements were done in water, whereas soil measurements were rare. This becomes obvious from the numbers of hits (in brackets at the y-axis). In summation, there were 1220 concentration values for rivers and lakes and 317 for corresponding sediments, while less than 100 were found for corn, 58 for fruits, and 168 for legumes. Even fewer measurements were found for soils. However, the measured values of PAHs and HCB

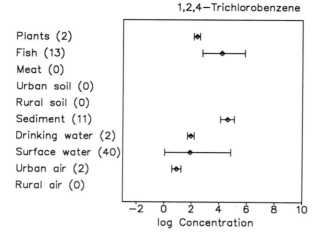

FIGURE 7.   Concentration ranges of 1,2,4-trichlorobenzene.

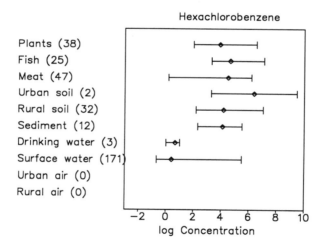

FIGURE 8.   Concentration ranges of hexachlorobenzene.

in fish, meat, and plants are typically 100 to more than 1000 times higher than the mean concentrations for water.

It is interesting to note that the allowed maximum content values established by many governments is much lower for water than for other food components. Table 1 shows the legal standards for a variety of toxic chemicals established by Germany (rsp. the European Community) and by the U.S. in various media. Most standards are for pesticides, with values for other chemicals being rare. Note that, due to a European edict, pesticide concentrations are

Table 1. Legal standards in drinking water and food in the European Community (EC) and in the United States (US)

	Water (mg/l)		Meat (mg/kg)		Plants (mg/kg)	
	EC[a]	US	EC	US	EC	US
CCl$_4$	0.003	0.005[b]			0.01–0.1	
Aldicarb	0.0001	0.003[b]	0.01	0.01[d]	0.1–0.5	0.05–1.0[d]
DDT	0.0001		1.0[f]	5.0 (fish)[c]	0.05–1.0	0.1–3.0[c]
Endosulfan	0.0001		0.1[f]	0.2[d]	0.1–30[g]	0.1–2.0[d]
Hexachlorobenzene	0.0001	0.001[e]	0.2[f]	0.3[c]	0.01–0.1	0.05–1.0[c]
Lindane	0.0001	0.0002[b]	1.0–2.0[f]	4.0–7.0[d]	0.1–2.0	0.01–3.0[d]
Mirex	0.0001		0.1[f]	0.1[c]	0.01	
Simazine	0.0001	0.004[e]	0.1	0.02[d]	0.1–1.0	0.2–12.0[d]

Source: European standards, Rippen[7].

[a] In the European Community: maximum pesticide concentration in drinking water is 0.0001 mg/l; sum of all is maximally 0.0005 mg/l.

[b] Reference 3.

[c] Reference 4.

[d] Reference 5.

[e] Reference 6.

[f] In fat.

[g] Maize 0.2 mg/kg, fruits and legume 1.0 mg/kg, hops 10 mg/kg, tea 30 mg/kg.

limited to 0.0001 mg/l (sum 0.0005) in drinking water. This limit is a precautionary value and does not necessarily represent a correlation with toxicity. The results shown in this table illustrate a paradoxical situation; i.e., limits in food are 300 to 1000 times higher than allowable limits in water.

Given the disparity between the standards (Table 1) and the higher concentrations found in foods (Figures 1 to 8), it is strange that the number of measurements for water content is more numerous than for food. Since the relative significance of anthropogenic chemical contamination in food is much higher than in water, it seems odd that the study of water and the regulatory limits for water have obviously taken precedence.

Historically, it is obvious that polluted water may carry various organisms responsible for serious health problems. For instance, the cholera epidemics in London and Hamburg at the end of the 19th century were associated with contaminated drinking water. Since water is so important to human health, food production, recreation, and transportation, it is natural to be concerned about its purity. However, since plants are central to all energy fixation and the basis for all food chains, it seems equally important to recognize that plant contamination has received far less attention.

In this regard several arguments deserve our attention. First, since water is often the source of food contamination, regulating water content may be the means for keeping food in check. Water is an easy standard that provides a consistent substrate, is found almost everywhere, and is the vector responsible for moving many chemical contaminants in the environment. Water is easy to sample and analyze, and since it is well mixed, statistical treatment of the data is uncomplicated. Problems associated with the lack of homogeneity, food selection habits, and food distribution, etc. are not associated with water. Alternately, food concentrations are typically higher than those of water, and consumption is as important and universal as for water.

Hopefully, the significance of anthropogenic chemicals in terrestrial ecosystems has been demonstrated with these few figures and tables. Some other considerations underline this argument: most of all biomass is found in terrestrial systems, and by far most of it is phytomass (Table 2).

Since phytomass constitutes such a large portion of the total biomass, and since plants are the base for all high-energy carbon molecules, it seems obvious that contamination of plants will have broad and dramatic consequences. Even a child easily understands this concept, but often, with the sophistication of age, many people forget this basic principle and dwell on small portions of ecosystem contamination while losing sight of the relative importance of plants.

Man is a terrestrial organism, part of the terrestrial ecosystem and dependent on the system for health and well-being. When the system is distorted, man cannot escape the consequences.

Terrestrial plants are contaminated in various ways with xenobiotic substances. The hazard is especially large when plants are growing on polluted sites. This includes waste sites and areas where contaminated sewage sludge has been applied. Another risk is the uptake of pollutants from air, because plants are specifically adapted to uptake of airborne substances. However, the number of compounds to consider is huge: up to 100,000 chemicals are traded. Among them, 3671 are of industrial importance; that means they are produced in large amounts. A total of 1816 compounds was found in the environment; rsp. their occurrence is very likely.[9] Graedel et al.[10] even talk of 2000 to 3000 substances that have been identified in the atmosphere alone, including metabolites of originally produced compounds.

Even much higher is the number of plant species, and also environmental influences are manyfold. Therefore, it seems to be impossible to determine by experiments alone the uptake and fate of any single chemical into any plant species under a certain set of environmental conditions. Tools are needed for the interpretation and prediction of the fate of chemicals in plants.

This book focuses on the processes that contribute to the uptake, accumulation, and elimination of anthropogenic organic chemicals by plants. Many of the factors that determine these processes are now understood in a biophysical sense. This means that from general physical laws much can be deduced about the problem of uptake of chemicals in plants. By using these laws in conjunction with principles of plant anatomy and physiology, models can be developed and predictions can be made about uptake and fate of pollutants. This includes the prediction of the importance of different uptake pathways, the estimation of kinetics, and the computation of equilibrium states for various chemicals and plants.

Table 2. Biomass (dry weight) on earth[8]

Mass (Pg = 10^{15} g)	Continents	Oceans
Phytomass	1837.0	3.9
Zoomass	1.0	1.0
Mankind	<.1	
Total biomass	1838.0	4.9

In applying these models one must keep in mind that a theoretical plant, although described on sound principles and true representations, does not accurately represent plants in natural settings. Features that allow a plant to successfully fill a particular niche are typically features that are not captured in simplistic models. Canopy structure, soil characteristics, natural environment, and cultivation practices also must be considered. However, despite the diversity of environment and the diversity of species, there are many commonalities in plants and their functions that are understood and are useful in making predictions about potential contamination. The challenge is to use these precepts and add knowledge about diversity in a manner that can yield realistic predictions in various crops and natural ecosystems.

V. ACKNOWLEDGMENT

Many thanks to Kristina Voigt for doing the study about the concentration ranges in the environment and all her coworkers from the "Information System on Environmental Chemicals" in the Project Group on Hazard Assessment of Environmental Chemicals, Munich-Neuherberg, Germany.

REFERENCES

1. Carson, Rachel (1962), *Silent Spring*, Houghton Mifflin, Boston.
2. Voigt, K., Pepping, T., Kotchetova, E., and Mücke, W. (1992), Testing of online databases in the information system for environmental chemicals with a test set of 68 chemicals. *Chemosphere*, 24: 857–866.
3. National Primary Drinking Water Regulations (1991), Part 150-189, Code of Federal Regulations (40 CFR) Ch.1 (7-1-91 Edition), Pesticide Tolerance Chemical/Commodity Index, pp. 515–662.
4. Pesticide Residues, FDA's Criteria for Enforcement Actions (1989), Guide 7141.01.
5. Pesticide Tolerance Chemical/Commodity Index (1991), Code of Federal Regulations (40 CFR) Ch.1 (7-1-91 Edition).
6. Current and Proposed National Primary and Secondary Drinking Water Regulations, U.S. Environmental Protection Agency Region X – Drinking Water Program, Latest revisions, 24 Apr. 1992.

7. Rippen, G. (1992), *Handbuch Umweltchemikalien (Handbook of Environmental Chemicals),* ecomed, Landsberg am Lech, Germany.

8. Sitte, P., Ziegler, H., Ehrendorfer, F., and Bresinsky, H. (1991), *Lehrbuch der Botanik für Hochschulen (Textbook of Botany for Universities),* 33rd ed., Gustav Fischer, Stuttgart, Germany, p. 251

9. Gesellschaft Deutscher Chemiker, Eds. (1988), *Altstoffbeurteilung (Judgment of Existing Chemicals),* GDCH, Frankfurt, Germany.

10. Graedel, T.E., Hawkins, D.F., and Claxton, L.D. (1986), *Atmospheric Chemical Compounds,* Academic Press, London, U.K.

Part One
Physiological

CHAPTER **2**

Anatomy and Physiology of Plant Conductive Systems

J. Craig Mc Farlane

TABLE OF CONTENTS

I. INTRODUCTION

    Mathematical models considered in this book are representations of the physical features and chemical reactions that define interactions between plants and their environment. By centering attention on equations, it is easy to lose sight of the intricate and complex nature of the problem. In this chapter I describe the anatomy of important plant features and briefly discuss some physiological principles that will help the reader to visualize and perceive the conditions which are represented in the models. I will draw attention to the physical structure of plants, but remind you that plants are living organisms which are successful because of biological (living) functions. These complex interactions of chemistry and physics which constitute life are incompletely understood. By presenting this discussion and the associated pictures, I hope to illustrate some of the assumptions and generalizations implicit in the mathematical approach to simulation.

The success of terrestrial plants is based on their peculiar ability to exist in an environment of dilute resources. Even in agricultural systems where water and nutrients are added and competing or herbivorous organisms are poisoned, plants must concentrate and conserve resources to live. The same features that account for accumulation and concentration of dilute resources also predispose plants to the accumulation of some anthropogenic chemicals. This allows the use of certain chemicals for successful pest control, while others become problematic pollutants.

Water is a key to much that happens in plants. The transpirational water flux serves as a conductive fluid, delivering mineral nutrients from dilute sources in the soil to where they are utilized in roots, stems, leaves, and fruits. Water is the solvent for the photosynthate moving in the phloem from leaves to stems, roots, and fruits. Water is a substrate for many biochemical reactions, is responsible for turgidity of tissues and stomatal control, and is an excellent coolant. Water also serves as a solvent, transport medium, and reactant for the uptake, translocation, and degradation of anthropogenic chemicals in the soil/plant environment. In total, water makes up between 70% and 90% of a typical actively growing herbaceous plant. Thus, models useful in describing pollutant considerations in plants must successfully describe water kinetics.

Because of the many competing interactions, the fate of chemicals in the soil/plant/air environment is not obvious. Models were thus developed to intelligently integrate available knowledge, to increase understanding of the complex interactions, to aid in presentation of plant functions, and to help make predictions about chemical fate.

II. ROOTS

A. STRUCTURE

Roots provide **anchorage, permit storage** of energy-rich molecules, and undergo a physical as well as a chemical **interaction with the soil**. A casual observation of plants dramatizes the variety in aerial growth forms but suggests little about the diversity below ground. However, as with shoots, roots also differ between species and environmental conditions (Lichtenegger and Kutschera-Nitter, 1991). Although not universally true, it is useful, and in many cases fairly accurate, to imagine as much of a plant below ground as above ground. Temporal distribution patterns are also important since roots grow into new areas of the soil and are active or become suberized as time proceeds. Figure 1 illustrates several root patterns and is presented as a reminder that plants are as different underground as in the forms we see.

Anchorage is generally proportional to the size and shape of the aboveground plant, yet this generalization has many exceptions. Redwood trees (the tallest and most massive terrestrial plants), for instance, are notably shallow rooted and depend on being associated with other trees for support against wind. Environmental conditions also influence rooting habits, such that plants in well-watered loam soils often will have shorter roots than the same plants in sandy, drier soils. Grasses in the prairie will have deeper roots than in your lawn or when potted and growing in a glasshouse. In modeling it is important to know rooting habits since position of roots over time determines exposure to anthropogenic chemicals and, thus, contamination.

The carrot demonstrates a feature that should be considered when developing models: "all is not as it appears". When the taproot of a carrot is harvested, there remains in the soil an expansive network of very fine roots which have supplied water and nutrients to the carrot plant. Because roots are not obvious many of us forget their existence and, therefore, their function.

The variety in root development patterns depends on the genetically controlled branching characteristics of the species. Differences also depend on radial growth, development of secondary cell walls, layering of suberin, and the deposition of stored materials. Branch roots anchor plants to the soil so that cells created at the terminal meristem are forced to expand through the soil. The root cap (Figure 2) is a layer of cells which develops distal to the apex and provides a cushion that protects the sensitive meristem. These cells slough off as they tear against soil particles but are continually replaced as the root grows. When a root confronts a barrier such as a rock, it continues to grow, following the path of least resistance until it is again free to grow downward. Root growth provides a continually changing source of nutrients, water, and possibly contaminants.

A short distance from the root apex many plants develop one-celled extensions of epidermal cells (root hairs) that dramatically increase the surface area of roots and make intimate contact with the soil (Figure 3). These root hairs are ephemeral, generally lasting only a short time, and are absent or haven't been observed in some species. Root hairs coincide with the zone of maximum water and cation uptake and have some control of plant desiccation in that they die when the soil dries. This zone is also where most anthropogenic chemicals enter plants. Since this zone is continually moving, as the roots grow into new areas of the soil, root dynamics are an important part of modeling.

Most crop growth models include root growth calculations (i.e., Sinclair, 1986; Amir and Sinclair, 1991). Others exist for forest (Dixon et al., 1990) and range plants. As in nature, models for root growth should include the influence of plant species (cultivar), soil characteristics, available nutrients, water, pests, meteorology, and solar radiation. Although root distribution is typically described on the basis of mass in a vertical direction, this information lacks details about root activity which may be critical to understanding contamination potentials.

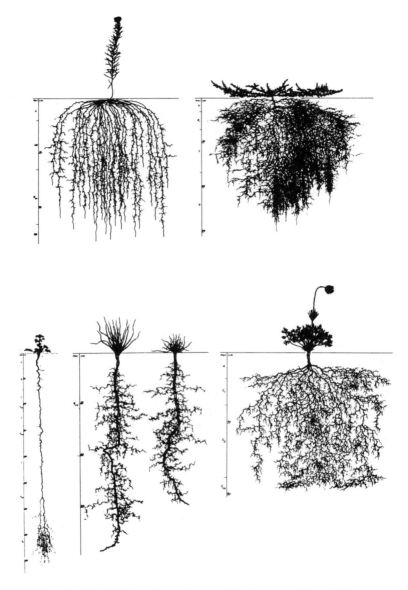

FIGURE 1. Root distribution patterns for a variety of species. (From Lichtennegger, E. and L. Kutschera-Nitter. 1991. in *Developments in Agricultural and Managed-Forest Ecology,* Vol. 24, **Plant Roots and Their Environment**, Proceedings of an ISRR Symposium, August 21–26, 1988, Uppsala, Sweden. McMichael, B.L. and H. Persson, Eds., Elsevier, New York, p. 360. With permission.)

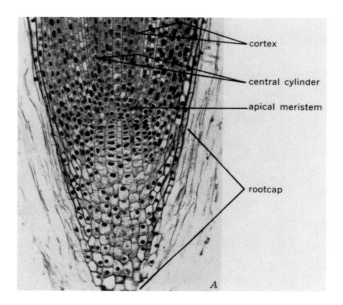

FIGURE 2.   Longitudinal section of a root showing the root cap, apical meristem, central vascular cylinder, and cortex. (From Esau, K., 1958, *Plant Anatomy,* John Wiley & Sons, New York, p. 630. With permission.)

Roots actively modify their environment. The **rhizosphere** is the zone in which colonies of bacteria, ascomycetes, and other fungi are nourished by organic nutrients exuded from roots. Some microorganisms contribute to plant health by modifying soil acidity, adding chelating agents, producing antigens to ward off pathogens, and expanding the effective absorption area, resulting in increased water and nutrient uptake.

Almost all plants are associated with fungi which may form a symbiotic relationship with the roots, the plant supplying energy-rich organic solutes and the fungus supplying mineral nutrients. These **mycorrhizae** (fungi plus root; Allen, 1991) were first classified by the manner of the connection between fungus and root. The **ecto**mycorrhizae are characterized by a mantle (mycelia forming around the root) or a hartig net (mycelia forming between cortical cells). The **endo**mycorrhizae are characterized by the presence of arbuscules or peletons which form within the cortical cells (but do not penetrate the plasmalemma as is often assumed). The small size (2 to 5 μm diameter) of the mycorrhizal fungi allows the hyphae to reach the very fine soil pores and thus acquire resources otherwise unavailable. Mycorrhizae have also been shown to increase water availability and to translocate almost any nutrient element from soil to plants.

Orchid seeds often will not develop in the absence of a fungal symbiont which supplies organic nutrients during specific periods of their growth. Hyphae

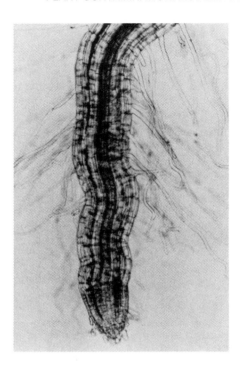

FIGURE 3.   Root of bent grass (*Agrostis tenuis*) showing root hairs which may be as long as 1.3 cm. Root hairs are comparatively short lived, dying as the root tip grows into new soil and the cortical and epidermal cells become suberized. (From Raven, P.H., R.F. Evert, and S.E. Eichhorn. 1992. *Biology of Plants*, 5th ed. Worth Publishers, New York, p. 478. With permission.)

of a VA mycorrhiza (*Glomus*) have been shown to remove ^{14}C-atrazine from soil and transfer it to corn (Nelson and Khan, 1990). These examples of organic chemical flux between fungus and plant are evidence that such interactions are possible; however, the importance and extent of fungus-mediated anthropogenic chemical uptake are not understood. It is clear that mycorrhizae are affected by various pesticides (and presumably industrial toxicants), but the effect differs between species and is generally less than the effect on the host plant (Trappe et al., 1984).

In terms of modeling plant contamination, there is currently insufficient information regarding the mycorrhizal influence on uptake and chemical metabolism to make this possible. However, there is sufficient inference to suggest that this is an area of potentially fruitful research.

B. PHYSIOLOGY

The number, arrangement, and size of various cell types differ between species, although the functions and features are common. A stylized root cross

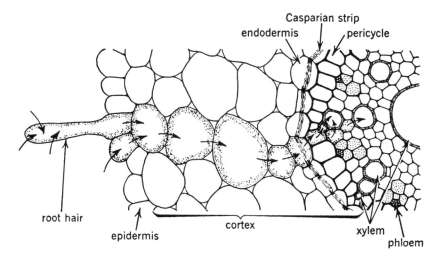

FIGURE 4.   Wheat root cross section showing possible routes of water and solute movement into the plant. (From Esau, K. 1958. *Plant Anatomy*. John Wiley & Sons, New York, p. 506.)

section is illustrated in Figure 4. Cell walls between the cortex cells are porous, often with voids where several cells adjoin. Water and solutes move freely from soil solution to the interior of roots in the capillary spaces between these cells. At the endodermis this movement is stopped by a barrier of waxy material, the **Casparian strip**, that is formed around the anticlinal walls of these specialized cells. The area outside the endodermis in which free water and solute movement occurs was recognized long ago by scientists studying mineral nutrition and was termed the **apparent free space**. This area also plays an important role in the movement of anthropogenic chemicals since this zone presents an extensive surface for the adsorption (partitioning) of pollutants.

At the endodermis all materials must pass through at least one cell membrane, thus entering the symplast. Membrane penetration of nutrient cations depends on metabolic energy and is termed "active uptake". Evidence of this process is demonstrated in Figure 5 where potassium concentration in solution remained constant when photosynthesis was suppressed by withholding CO_2. Under these conditions water movement was unaffected, indicating that its uptake and movement were passive, driven by the water potential gradient that results primarily from the evaporative loss by leaves. With the exception of some hormone-like chemicals (2,4-dichlorophenoxyacetic acid [2,4-D]) there is **no** evidence of **active uptake** for anthropogenic chemicals as they move through the endodermis and into the vascular system of plants. Rather it has been shown to be a passive process controlled by diffusion, solubility in water, and solubility in the membrane (see discussion by Trapp in Chapter 5 and Bromilow and Chamberlain in Chapter 3).

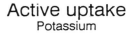

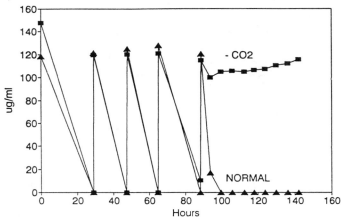

FIGURE 5. Potassium (K^+) decreased in the hydroponic solution (make-up solution was H_2O) of two adjacent airtight growth chambers. After the third day, the CO_2 injection was turned off in one of the chambers and the procedure of sampling, draining, and replacing the nutrients was repeated for an additional 2 days. At the end of the second day, cation uptake was completely stopped while the control plants continued to function the same throughout the test.

III. STEMS

Stems of trees and many crop plants are visually obvious, yet the stems of other plants, like lettuce or carrot, are inconspicuous. Stems typically support leaves in an arrangement that provides efficient collection of solar radiation and gaseous exchange with the atmosphere. Flowers and fruits are also supported on stems, and in some plants, stems constitute the main site for storage of carbon. Tubers, such as potatoes, are morphologically stems, and although they reside underground and are thus exposed to the soil, they function as stems.

Stem length becomes more important in modeling as it increases. In tall trees the distance from roots to leaves may be the dominant factor that controls which chemicals reach the leaves. Partitioning of chemicals to materials in the stem increases with increasing K_{ow} values (McCrady et al., 1987), and separation of chemicals follows the principles that allow chromatographic separation for analysis. Although many cells in plants are dead, live cells, especially near the cambium and phloem, may provide an area for rapid metabolic degradation of anthropogenic chemicals. Very little is known about this feature which should be the subject of significant research.

IV. LEAVES

In some species leaves are rigid, and in others leaf orientation changes depending on solar intensity, water status, and time of day or year. Leaves avoid overheating by having good radiative properties and an axillary cooling system based on water evaporation which is regulated by the degree of stomatal opening and availability of water. The leaves are covered with protective cuticles that function primarily to retard water loss and also protect the plant from infection by numerous pathogens. The cuticle is a complex structure consisting of a **pectin** layer, which bonds the **cutin** to the cell walls (Figure 6), and a layer of **epicuticular wax** on the outside (Figure 7). The fine structure of the wax layer varies from simple and thin to thick and complex and differs between species (Figure 8). The waxes are complex mixtures containing long-chain hydrocarbons, alcohols, ketones, fatty and hydroxy-fatty acids, and esters; some waxes also contain appreciable proportions of aldehydes (Martin and Juniper, 1970). Natural pores (holes) in the cuticle (Figure 9; Mc Farlane, 1971) account for some transpiration and the penetration of some nutrients and anthropogenic chemicals. The number and size of these pores change as plants age and weather. (See Chapter 6 for details.)

The **hydrophobic** cuticle precludes the penetration and, thus, the collection of CO_2 by underlying cells. However, when the stomata are open, gas molecules diffuse in and out and interact with a large **hydrophilic** area of water-covered cells (Figure 10). This surface is responsible for plant accumulation of CO_2 and loss of O_2 and H_2O vapor. When stomata are closed, hydrophilic pollutants such as O_3 and SO_2 are ineffective in causing plant injury, but when they are open, damage is proportional to concentration and duration of exposure. On the other hand, some of the most problematic pollutants are very lipophilic, and the large lipid-covered surfaces of leaves (cuticles) form an ideal sink for accumulation of these chemicals. For these chemicals, stomatal opening has little or no effect on the processes of contamination and pollutant loss from leaves.

On the surface of many leaves are hairlike structures (Figure 11) that greatly increase the air contact surface. On some plants they effectively increase the air-to-surface boundary layer, creating a stagnant area which decreases transpiration rates. On other plants the hairs contain various materials that may discourage or attract insects. Most of us have experienced these trichomes, which account for the characteristic odor and yellow material that covers your hands when you are working with tomato plants and the unpleasant experience of contacting stinging nettle. The physiological functions vary and in many cases are unknown. With regard to contamination by anthropogenic chemicals, these structures increase air/surface contact and present a variety of substances for yet unknown chemical interactions.

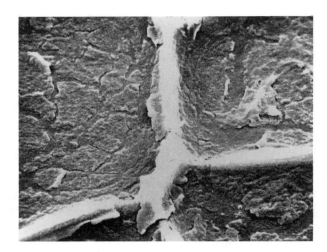

FIGURE 6.   Inside surface of top (no stomata) cuticle from apricot leaf. This cuticle was separated from the leaf with $ZnCl_2$ HCl digest. The pectin layer is seen as a cracked layer upon the cutin. (From Mc Farlane, J.C., 1971, Cuticular Permeability to Mineral Nutrients, Ph.D. dissertation, University of California, Riverside, p. 49. With permission.)

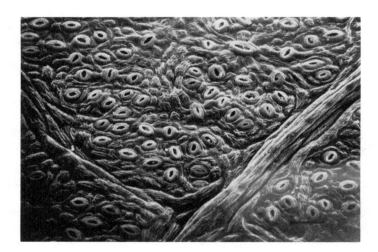

FIGURE 7.   Bottom surface of apricot leaf cuticle showing relief and position of stomata.

From the discussion above it is obvious that plants are well adapted to accumulate airborne pollutants. Countering this vulnerability is the compensating feature of leaves which are generally oriented for efficient interception of solar radiation. This orientation also provides maximum exposure to ultraviolet

A

B

FIGURE 8.   Epicuticular wax of (A) *Zea mays* leaf and (B) *Pinus sylvestris needle.*
Electron micrographs made of carbon replicas. Cr/Au/Pb. (From Martin, J.T.
and B.E. Juniper, 1970, *The Cuticles of Plants,* St. Martin's Press, New
York, pp. 102 and 110. With permission.)

radiation and thus optimizes the possibility of photodegradation of pollutants.
This has been demonstrated to be an important factor in the persistence of dioxins
on plants (McCrady et al., 1987). A large water flux may cause some chemicals
to move into plants; however, in many cases that same flux facilitates movement

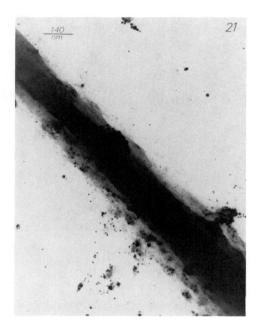

FIGURE 9. Electron-dense spots indicate pores in the cuticle of apricot leaves above and adjacent to the anticlinal walls of adjacent epidermal cells. The AgCl precipate was formed in the pores by putting a solution of $AgNO_3$ on one side of the cuticle and HCl on the other side. (From Mc Farlane, J.C. 1970. Cuticular Permeability to Mineral Nutrients, Ph.D. dissertation, University of California, Riverside. With permission.)

of chemicals to the substomatal tissues and subsequent loss via volatilization (Mc Farlane and Pfleeger, 1990).

V. VASCULAR TISSUE

Xylem and **phloem** differentiate near the root apex (Figure 12) and form a continuous network throughout the plant, terminating within several cells of the stomata (Figure 13). The movement of water, nutrients, energy-rich photosynthate, and also pollutants occurs primarily in these tissues.

A. XYLEM

The xylem includes fibrous cells, which provide support; parenchyma cells, which often function as storage areas for starch and fat; and the main

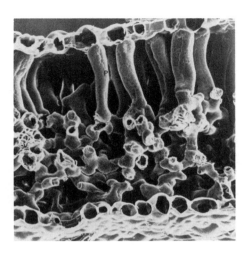

FIGURE 10.   Transverse view of a broad bean leaf (*Vicia faba* L.). Beneath the upper epidermis is a layer of vertically elongated palisade mesophyll cells (**P**). Below is a loosely arranged layer of spongy mesophyll cells with considerable air space between them and a vein (**V**) containing the xylem and phloem. (From Troughton, J.H. and F.B. Sampson. 1973. *Plants, A Scanning Electron Microscope Survey*. John Wiley & Sons, New York, p. 120. With permission).

functional cells, which are the "tracheary elements". This is a term applied to cells that resemble the tracheae of animals and have been shown to be responsible for most of the water movement in plants. Ironically, when fully functional, these cells have lost their membranes and protoplast and are thus dead. In gymnosperms and some lower vascular plants, the conduction cells of the xylem are named **tracheids**. They are elongated (some reaching 2 to 7 mm) with the end walls intact (Figure 14) and are characterized by openings between adjacent cells (pit pairs) that form a continuous open pathway when fully developed. The pits occur on all surfaces, creating a network for water movement along the cell length with easy connection to adjacent cells (Figure 15). In the spring, when water, nutrients, and weather favor growth, the diameter of first-formed tracheids (springwood) is greater than the diameter of later-developed tracheids (summerwood). Thus, in perennial plants, a visible ring is formed in the stem each growing season.

Other plants (angiosperms) have **vessel elements** in which the end walls collapse, forming open continuous pathways for water and solute movement (Figure 16). They are also connected to adjacent cells by pitted walls, but it is easily understood from a simple observation that most of the vertical water movement occurs in the vessels.

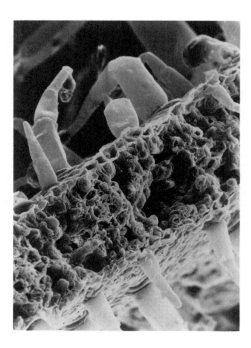

FIGURE 11. Tobacco (*Nicotiana tabacum*) leaf transverse section. The trichomes (**T**-hairs) are modified epidermal (**E**) cells on the top and bottom which greatly increase the contact surface with the air and, thus, gaseous pollutants. The mesophyll (**M**) cells are the site of CO_2 absorption and photosynthesis. The vascular (**V**) tissue includes both xylem and phloem. (From Lott, J.N.A. 1976. *A Scanning Electron Microscope Study of Green Plants.* C.V. Mosby, Saint Louis, MO, Plate 16, p. 16. With permisison.)

1. Water Flux

Although species differ in the number, size, and arrangement of tracheary elements, they all function in a similar manner in the conduction of water and solutes. When stomata are open and environmental conditions allow transpiration, a continuous water potential gradient is created throughout the plant. In the xylem the energy gradient is expressed as a pressure potential in which water and solutes move by mass flow, much like in a hose, where the flux responds to differences in pressure from one point to another. Dissimilar to the hose analogy, this pressure gradient is negative to the atmosphere, a condition that normally would restrict movement to a height of 1 atm (10.3 m), hardly the height of many trees. The forces of water molecular adhesion and surface interactions between water and cell wall materials provide the support the water needs to allow tree growth in excess of 100 m.

Water evaporation requires 2.5 MJ/l. On an average midsummer clear day at 40° north latitude, the incident radiation is approximately 30 MJ/(m² · day).

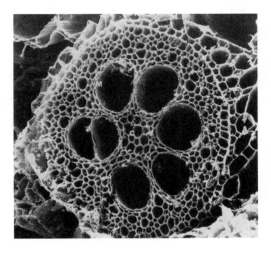

FIGURE 12. Transverse section of a corn root (*Zea mays*). The vascular tissue is surrounded by the endodermis (**E**) and the pericycle (**P**) immediately beneath. Vessels of the late metaxylem (**M**) are large, and the protoxylem (**C**) cells are in the center with pith cells surrounding the xylem. (From Troughton, J.H. and F.B. Sampson. 1973. *Plants, A Scanning Electron Microscope Survey.* John Wiley & Sons, New York, p. 111. With permission.)

If all that energy was used to evaporate water, the potential would be 30/2.5 = 12 l/(m^2 · day). Bugbee and Salisbury (1988) have measured values as high as 10 l/(m^2 · day) in hydroponic, unstressed wheat, and average values of 4.2 to 5.7 l/(m^2 · day) are reported for various crops (Table 1).

It is interesting to imagine water loss from a field or forest in common units. Assuming a value of 5 l/(m^2 · day), equivalent values are 5,346 gal/acre, or 50,000 l/ha daily. In a growth chamber with four 30-day-old (average 53 g shoot fresh weight) soybean plants, the transpiration rate [10 to 13 ml/(h · plant)] can be observed easily as the condensed water collects in a glass tube. This is equivalent to changing the entire water content of the shoots every 3 to 4 h.

Molecular diffusion occurs at all times and is defined by molecular weight, temperature, molecular free energy, and the nature of the medium. Water movement across membranes is properly described as diffusion even though an osmotic gradient exists favoring net movement of water in one direction. In this case the molecular free energy of the water molecules on the solute side of the membrane is reduced by hydrogen bonding with the solute or structural material of the container (cell walls), decreasing the average molecular **free** energy of water and thus reducing the reverse diffusion. The net result is observed as unidirectional water flux.

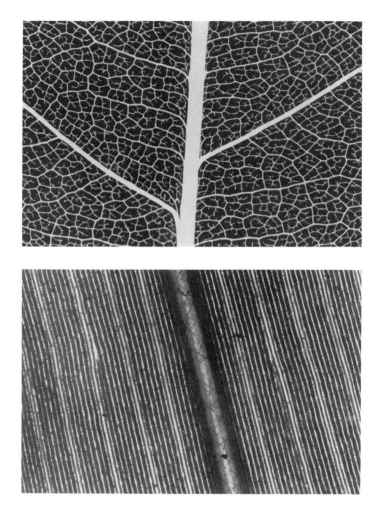

FIGURE 13. Parallel and net venation patterns of monocot (bottom) and dicot (top) plants demonstrate the intricate patterns and extent of the vascular system. (Photograph by Andreas Feininger, © 1977, from *Leaves,* Dover Publications, New York, pp. 44 and 45. With permission.)

Mass flow of water carries dissolved and suspended solutes needed for nutrition, but also some pollutants. Lateral movement of water and solutes to the tissues adjacent to the xylem also occurs by diffusion even while the bulk of the water is being drawn up the xylem. This lateral movement accounts for transfer of solutes to all adjacent tissues, with the degree of exchange being dependent on the tortuosity of the pathway and characteristics of the tissue in relation to the solute chemical. Since the phloem is typically close to the xylem,

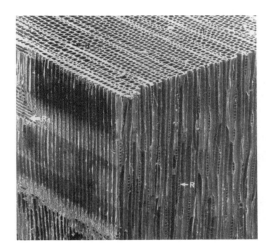

FIGURE 14.   Redwood (*Sequoia sempervirens*) wood showing elongated tracheids (**T**), which are the dead yet main functional cells in conducting water and solutes from roots to the leaves. The ray (**R**) cells provide connection between cells of different age connecting in a direction radial from the center of the stem. (From Troughton, J.H. and F.B. Sampson. 1973. *Plants, A Scanning Electron Microscope Survey*. John Wiley & Sons, New York, p. 77. With permission.)

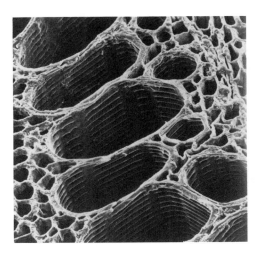

FIGURE 15.   Tracheids in fern showing elongated pits in the side walls arranged like a ladder; this is called scalariform pitting. (From Troughton, J.H. and F.B. Sampson. 1973. *Plants, A Scanning Electron Microscope Survey*. John Wiley & Sons, New York, p. 68. With permission.)

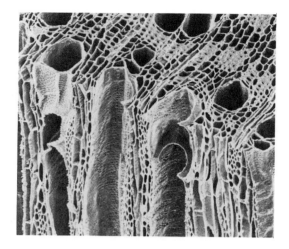

FIGURE 16. Wood of *Eugenia maire* showing the remnants of the vessel end walls which join together forming a continuous pipe for water movement. Notice the pitting in the vessels which provides openings for solution and solute movement to the other cells in the stem. (From Meylan, B.A. and B.G. Butterfield. 1978. *The Structure of New Zealand Woods.* Bulletin 222, DSIR, Wellington, New Zealand, p. 86. With permission.)

Table 1. Evapotranspiration (E/T) for several crops

Crop (l/m²)	E/T (Daily average)
Maize	4.9
Sorghum	5.3
Potato	4.2
Sugar beet	4.6
Wheat	4.2
Soybean	5.3
Alfalfa	5.7

Source: Jensen, 1973.

transfer of solute is expected, but since there is no mechanism for sequestering anthropogenic chemicals in the phloem sap, the concentration will not exceed the concentration in the xylem. (Bromilow and Chamberlain, 1989).

B. PHLOEM

The main conducting cells of the phloem are the sieve elements. They are distinguished by sieve areas, which consist of numerous pores giving the appearance of a sieve (Figure 17), but differ from the xylem cells (tracheary elements) in that they have functional membranes and are filled with living

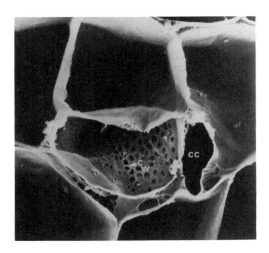

FIGURE 17. Sieve plate in the phloem of pumpkin stem (*Cucurbita pepo*). Sieve plate consists of cell wall (c_w) and numerous openings called pits (p). Companion cell (cc) is adjacent to phloem vessel. (From Troughton, J.H. and F.B. Sampson. 1973. *Plants, A Scanning Electron Microscope Survey.* John Wiley & Sons, New York, p. 100, plate 108. With permission.)

protoplasm. In angiosperms the sieve plates are on end walls connecting a series of cells together to form sieve tubes. In gymnosperms and most lower plants they follow the pattern of the tracheary elements in that they are obliquely joined or have the sieve plates on side walls. Nevertheless, the function in all plant types is similar. The protoplast of mature sieve cells is peculiar in that no intact nuclei and no vacuole can be distinguished and all membranes are adjacent to the cell walls. Adjoining sieve cells are characteristic cells called companion cells (Figure 18). They are closely connected to the sieve elements by numerous plasmodesmata (small protoplasmic connections similar to those of the sieve plates) and function with the sieve elements by supplying the energy-rich molecules.

Solution flow in the phloem has been the object of numerous theories. The most accepted hypothesis (currently) depends on mass flow of solution created by a gradient of osmotic pressure along the path of the phloem. This source/sink concept depends on phloem loading at the source. In mature plants the source of assimilate is mostly sugar which is formed by photosynthesis in the leaves. In seedlings the source is the cotyledons, and in periods of stress the source may be fruits or stored material in the stem or roots. The solutes (more than 90% of phloem solute is typically sucrose) are moved into the phloem cells by an active process which requires energy in the form of ATP. Once in the cells, the solutes create an osmotic gradient favorable to the entrance of water. At the sink, the sugars leave the phloem and are generally altered to form

COMPANION   SIEVE   50 μm
    CELL   PLATE
              (e)

FIGURE 18.   Phloem cells are filled with protoplasm and connected primarily at the ends by sieve plates. Companion cells are involved with the transfer of solutes into and from the sieve elements for transport. The arrows point to P-protein bodies in immature cells. (From Raven, P.H., R.F. Evert, and S.E. Eichhorn. 1992. *Biology of Plants,* 5th ed., Worth Publishers, New York, p. 463. With permission.

starches or complex sugars. This changes the osmotic condition in the phloem and creates a situation favorable for the loss of water. Water loss at the sink creates a pressure gradient in the phloem that causes the sap (phloem solution) to move by mass flow, carrying with it the solutes from one location to another. Flow rates in phloem vary from a few centimeters to more than 100 cm/h.

VI. CONCLUSIONS

Many studies in plant physiology are conducted with seedlings; this often yields useful inferences to features in mature plants. However, when considering water flow and energy-dependent phenomena such as those that determine the uptake and distribution of anthropogenic chemicals in plants, considerable caution should be observed. When many seeds germinate, most of the early development occurs in roots, with the above-ground organs enlarging only after a firm connection to the soil is established. Materials and energy are derived from the seed. Cell enlargement is based on carbon redistribution, and water is delivered by an osmotic gradient. Photosynthesis and transpiration are often very limited, and the biochemical reactions differ dramatically from those of mature plants. This should generate considerable caution to modelers when considering seedling experiments for verification of model results.

Although this chapter has pointed to several aspects of plants that contribute to the effective collection of certain pollutants, there are other features that provide a degree of protection. These include the following:

- Edible portions of plants typically result from phloem transport of the building materials (sugar, amino acids) which are polar and have specific mechanisms that facilitate movement into and from the phloem. Lipids, proteins, and other compounds which are not easily transported in the phloem are primarily synthesized within the storage tissue. Transport of toxic materials in the phloem is generally limited by the nature of the chemical, which is often incompatible with membrane transport.
- Plant cells have a complement of very active enzymes and, thus, metabolic degradation of many anthropogenic chemicals is rapid.
- Sorbens in soils often limit chemical solubility and, thus, availability to plant uptake.
- Leaf orientation facilitates photodegradation of chemicals that are adsorbed to the surface.
- The rapid flux of water through plants tends to flush pollutants to areas where volatile chemicals may volatilize and thus leave the plant.

Our current understanding of the anatomy and physiology of plants is sufficiently complete to allow mathematical models to represent chemical uptake from soil and atmosphere and chemical fate within plants. In the quest to expand the utility of models, it is important to remember the intricate nature of the anatomy, morphology, and biochemical reactions that make plants the most productive and resilient organisms on earth. Modeling anthropogenic organic chemicals in plants will lead to a better understanding of the fundamental character of plants, aid pesticide design and application, and help in exposure assessments of pollutant contamination of food chains.

REFERENCES

Allen, M.F., 1991, *The Ecology of Mycorrhizae*, Cambridge University Press, New York.

Amir, J. and T.R. Sinclair, 1991, A model of the water limitation on spring wheat growth and yield, *Field Crops Res.*, 28:59–69.

Bromilow, R.H. and K. Chamberlain, 1989, Mechanism and regulation of transport processes, in Designing Molecules for Systemicity, British Plant Growth Regulatory Group, Monograph 18:113–128.

Bugbee, B.G. and F.B. Salisbury, 1988, Exploring the limits of crop productivity, I. Photosynthetic efficiency of wheat in high irradiance environments, *Plant Physiol.,* 88:869–878.

Dixon, R.K., R.S. Meldahl, G.A. Ruark, and W.G. Warren, Ed. 1990, *Process Modeling of Forest Growth Responses to Environmental Stress*, Timber Press, Portland, OR.

Esau, K., 1958, *Plant Anatomy,* John Wiley & Sons, New York.

Feininger, A., 1984, *Leaves,* Dover Publications, Mineola, NY.

Jensen, M.E., 1973, *Consumptive Use of Water and Irrigation Water Requirements,* American Society of Civil Engineers, New York.

Lichtenegger, E. and L. Kutschera-Nitter, 1991, Spatial root types, in *Developments in Agricultural and Managed-Forest Ecology,* Vol. 24, Plant Roots and Their Environment, Proceedings of an ISRR Symposium, August 21–26, 1988, Uppsala, Sweden. McMichael, B.L. and H. Persson, Eds. New York, pp. 359–365. Elsevier.

Lott, J.N.A., 1976, *A Scanning Electron Microscope Study of Green Plants,* C.V. Mosby, Saint Louis, MO.

Martin, J.T. and B.E. Juniper, 1970, *The Cuticles of Plants,* St. Martin's Press, New York.

McCrady, J.K., C. Mc Farlane, and F.T. Lindstrom, 1987, The transport and affinity of substituted benzenes in soybean stems, *J. Exp. Bot.,* 38(196):1875–1890.

Mc Farlane, C. and T. Pfleeger, 1990, Effect, uptake and disposition of nitrobenzene in several terrestrial plants, *Environ. Toxicol. and Chem.,* 9:513–520.

Mc Farlane, J.C., 1971, Cuticular Permeability to Mineral Nutrients, Ph.D. dissertation, University of California, Riverside.

Meylan, B.A., and B.G. Butterfield, 1978, *The Structure of New Zealand Woods.* Bulletin 222, DSIR Wellington, New Zealand.

Nelson, S.D. and S.U. Khan, 1990, The role of *Glomus* mycorrhizae in determining the fate of atrazine in soil and corn plants, Eighth North American Conference on Mycorrhizae, Jackson, WY, Sept. 1990.

Raven, P.H., R.F. Evert, and S.E. Eichhorn, 1986, *Biology of Plants,* Worth Publishers, New York.

Sinclair, R.R., 1986, Water and nitrogen limitations in soybean grain production. I. Model development, *Field Crops Res.,* 15:125–141.

Trappe, J.M., R. Molina, and M. Castellano, 1984, Reactions of mycorrhizal fungi and mycorrhiza formation to pesticides, *Annu. Rev. Phytopathol.,* 22:331–359.

Troughton, J.H. and F.B. Sampson, 1973, *Plants, A Scanning Electron Microscope Survey.* John Wiley & Sons, New York.

Part Two
Chemical

CHAPTER **3**

Principles Governing Uptake and Transport of Chemicals

Richard H. Bromilow and Keith Chamberlain

TABLE OF CONTENTS

1-56670-078-7/95/$0.00+$.50
© 1995 by CRC Press, Inc.

I. INTRODUCTION

In order to model the behavior of organic chemicals in plants, one has to understand quantitatively the fundamental processes that influence uptake, transport, and metabolism. From the present state of knowledge, it would appear that much behavior (apart from metabolism) can be satisfactorily explained, at least in general terms, from physicochemical properties such as acid strength and lipophilicity. Metabolism is more difficult to predict and, unlike the transport processes, may differ markedly among plant species.

Rate of passage of compounds across membranes is one of the crucial factors in transport, and, in this respect, the behavior of nonionized chemicals is easier to predict than that of ionized compounds. Particular difficulties arise with molecules existing in several different ionized forms, for little is known about the movement of complex anions or zwitterions across membranes. Phloem transport is difficult to predict for such compounds, and indeed only in recent years have predictions (more qualitative than quantitative) of phloem transport been possible even for nonionized or singly ionized molecules. This chapter attempts to summarize the available knowledge on the principles controlling the uptake and transport of chemicals, and also to indicate areas where predictions will be unreliable due to the lack of fundamental knowledge about the detailed processes.

II. INFLUENCE OF PLANT STRUCTURE ON THE UPTAKE AND TRANSPORT OF XENOBIOTICS

The structure of plants, with particular regard to the anatomy and physiology of the conducting tissues, has been discussed in the preceding chapter. A few of the more important points bear repetition here.

Compounds entering plant roots can move in the intercellular spaces of the cortex (part of the apoplast) until they reach the endodermis, a cylindrical sheath of cells that are tightly bound together by the lignified Casparian strip and which surrounds the vascular tissues in the stele. To proceed beyond the endodermis, compounds must enter the endodermal cells and can then move further into the vascular tissues. Thus, compounds taken up by roots and moved to shoots via the apoplastic system of the xylem nonetheless also must have the ability to enter the symplast.

A second point of note is the differing pH of the various compartments of plants, with xylem typically at about 5.0, phloem at 8.0, cell vacuoles at 5.5, and cytoplasm at 7.5. These pH differences do not influence the distribution of nonionized compounds between the compartments, but do strongly influence the distribution of ionized compounds; e.g., weak acids tend to accumulate in compartments of high pH (see Section III.D). Thus, only a restricted range of nonionized compounds are mobile in phloem, for most such compounds move freely between the adjacent phloem and xylem tissues and so tend to move with the far greater water flow in the xylem; however, many weak acids can be retained in phloem due to their substantial ionization at pH 8.0 and, hence, can be moved with the phloem sap to areas of new growth.

III. PHYSICOCHEMICAL PROPERTIES AND THEIR ROLE IN DETERMINING UPTAKE AND TRANSPORT

A. VAPOR PRESSURE

Movement from soil to plant via the vapor phase can be important for some compounds. Certain herbicides, such as trifluralin, are so volatile that sprays onto soil surface can give rather poor weed control due to the rapid loss by volatilization — shallow incorporation is often necessary to retain trifluralin in soil sufficiently for it to be an effective herbicide. Vapor pressures of pesticides are listed by Worthing and Hance,[1] Ashton and Crafts (herbicides only),[2] and Howard.[3] The data for industrial chemicals are more scattered; Howard[4] covers a range of bulk-produced chemicals, priority pollutants, and solvents, and Hutzinger[5] gives values for groups of halogenated solvents, halogenated aromatic compounds (including polychlorinated biphenyls [PCBs]), fluorocarbons, phthalates, and phosphate esters. Vapor movement of compounds in the soil-water-air system depends on their distribution between these three phases; this is discussed in more detail in Section IV.B.2.

B. LIPOPHILICITY AND WATER SOLUBILITY

The most important property controlling the movement of chemicals in plants is the lipophilicity, i.e., the balance between the affinity of the chemicals for aqueous phases and that for lipid-like phases. Lipophilicity determines the ease of movement across plant membranes and so, for example, determines the potential for long-distance transport via xylem or phloem. Other properties of chemicals, such as the presence of particular functional groups or the molecular weight (up to at least 450), appear to have little influence on systemic properties. Partitioning of compounds onto plant solids also limits long-distance transport, and this process, too, is a simple function of lipophilicity.

Lipophilicity is usually assessed using 1-octanol/water partition coefficients (K_{ow}). These can be measured using shake-flask methods[6] or high-pressure liquid chromatography (HPLC) on silica columns coated with 1-octanol. For log K_{ow} values greater than 4.0 these methods are difficult, and indirect methods are used such as reverse-phase HPLC on octadecylsilica columns.[7] Alternatively, predictions of log K_{ow} can be made using fragment constants,[8] but accurate estimates usually depend on having a measured baseline value for at least one compound of that particular structural type.

Water solubility is generally rather a poor guide to systemic behavior. However, lipophilicity can be estimated from water solubility by linear free-energy relationships such as those given by Briggs:[9]

$$\text{For liquids, log WS} = 0.84 - 1.18 \log K_{ow} \qquad (1)$$

$$\text{For solids, log WS} = 0.01 - \log K_{ow} - (0.01\, T_m - 0.25) \qquad (2)$$

where WS is the water solubility (mol/l) and T_m is the melting point in °C. The melting point term provides an estimate of the lattice energy of the crystal, and makes allowance for the fact that the property of water solubility of solids is a function of both lipophilicity and this lattice energy.

The ranges of log K_{ow} values for some common classes of organic chemicals found in the environment are shown in Figure 1. Classes of compounds such as the organophosphorus insecticides encompass a wide variety of structures, and consequently their log K_{ow} values span a wide range.

C. ACID STRENGTH

Certain compounds are appreciably ionized at the pH values found in soil or plant cells, and this complicates interpretation of their behavior. Ionized species are much more polar than the corresponding nonionized forms, and so their movement across plant membranes can be very different. For example, many herbicides are monobasic carboxylic acids having pKa values in the range 3 to 6, and so these compounds will be substantially ionized in soil and plants; the anions so produced are typically 3 to 4 log K_{ow} units more polar than the undissociated parent acids.

In measuring log K_{ow} for such ionizable compounds, care must be taken that these measurements are made at pH values at which essentially only one species is present; if this is not possible, then corrections must be applied. The pKa can be regarded most simply as the pH at which a particular acid or base group is 50% ionized, and pKa can be measured[10] or predicted[11] by standard procedures.

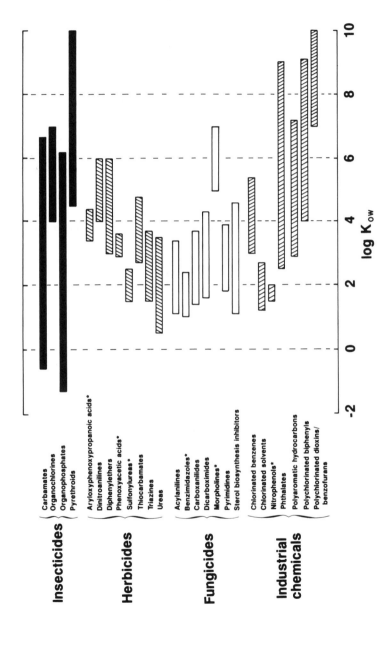

FIGURE 1. Ranges of 1-octanol/water partition coefficients (as log K_{ow}) for commonly occurring compounds in various classes of pesticides and industrial pollutants. Asterisks indicate ionizable compounds, whose log K_{ow} values are plotted for the undissociated molecule.

D. TRANSPORT ACROSS MEMBRANES AND ION TRAPPING

The movement of nonionized compounds across membranes appears to be largely a function of the lipophilicity of the compound, and in Section IV this is discussed in more detail under the relevant experiments. For ionizable compounds, however, the situation is more complex, especially where a membrane separates plant compartments of different pH (e.g., the plasmalemma, which encompasses the cell contents within the cell wall; and the tonoplast, which separates vacuoles, pH ~5.5, from cytoplasm, pH ~7.5). Under such circumstances, there will be different proportions of dissociated and undissociated forms in the two compartments. If one considers the situation for a weak acid, the proportion of undissociated compound will be highest in the compartment of low pH; since this form will cross the membrane more rapidly than the more polar corresponding anion, undissociated molecules moving from the low-pH compartment will be substantially ionized in the high-pH compartment, and the anions so formed will be unable to escape freely from the high-pH compartment. This process forms the basis of the "ion trap" effect, whereby weak acids are accumulated inside plant compartments of high pH provided that the compounds do not damage or overwhelm the proton pumping that maintains the pH values in these compartments.

The example in Figure 2 illustrates how a carboxylic acid of pKa 4.0 might be accumulated in a phloem sieve tube relative to the apoplast.[12] If it is assumed that the anion is not able to cross the membrane, then the ratio of total concentrations in the two compartments,

$$[(A^- + AH)_{inside}/(A^- + AH)_{outside}]$$

shows a large accumulation in the phloem — that is, $(10,000 + 1)/(10 + 1) = 909$. In practice, however, anions do leak across the plasmalemma membrane to some extent, and this limits accumulation. There is also a small effect of the membrane charge on the permeation of anions; Briggs et al.[13] have utilized the following equation[14] for accumulation which takes all these factors into account:

$$\frac{[AH]_i + [A^-]_i}{[AH]_o + [A^-]_o} =$$

$$\frac{\left(1 + 10^{pHi-pKa}\right)\left\{P_{AH}/P_{A^-} + \left[(FE/RT)/\left(1 - e^{-FE/RT}\right)\right]10^{pHo-pKa}\right\}}{\left(1 + 10^{pHo-pKa}\right)\left\{P_{AH}/P_{A^-} + \left[(FE/RT)/\left(1 - e^{-FE/RT}\right)\right]10^{pHi-pKa} \cdot e^{-FE/RT}\right\}} \quad (3)$$

where pHi and pHo are, respectively, the pH inside and outside of the membrane, F is the Faraday constant (96,487 C/mol), E is the membrane potential (V), R is the universal gas constant (8.31 J/K · mol), T is the absolute

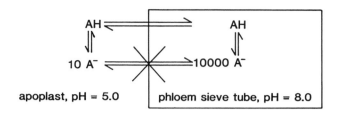

FIGURE 2. Accumulation of weak acids within cells by the ion-trap effect. See text for details. (From Bromilow, R.H. and Chamberlain, K. (1989), in *Mechanisms and Regulation of Transport Processes,* R.K. Atkin and D.R. Clifford, Eds., Monogr. 18, British Plant Growth Regulator Group, p. 113. With permission.)

temperature (K), and P_{AH} and P_{A^-} are the permeation rates of the undissociated and dissociated forms of the acid, respectively, through the membrane. The membrane potential of the plasmalemma is typically around -0.12 V, while that of the tonoplast is about two orders of magnitude smaller. Accumulation is greatest at highest values of the permeability ratio (P_{AH}/P_{A^-}).

IV. UPTAKE AND TRANSPORT PROCESSES FOR XENOBIOTICS

A. STUDY SYSTEMS

Uptake by the roots of whole plants and subsequent translocation with the transpiration stream in the xylem are relatively easy to study, and many plant species have been used. It is usual to apply the xenobiotic (normally radiolabeled) to the plant roots in nutrient solution and to measure the amounts accumulated in the different plant parts after a certain period of time. Concentrations of the xenobiotic can be measured in xylem sap collected from the stem or leaf bases, as demonstrated for several species by Mc Farlane et al.[15] using a vacuum tube and by Hsu et al.[16] for soybean *(Glycine max)* using a pressurized root chamber to give realistic flow rates from the stump of a cut stem. Alternatively, the time-averaged concentration of chemical in the xylem sap may be estimated from the amount of chemical accumulated in the plant shoots for a known volume of water transpired by the plant; correction must be applied if the compound is appreciably degraded in the plant over the period of the test.

For some lipophilic compounds, movement from soil or nutrient solution to foliage can occur primarily via the vapor phase rather than by systemic transport. Schroll and Scheunert[17] discussed experimental approaches to this problem and described a procedure to separate these two processes in a study utilizing hexachlorobenzene.

The study of phloem translocation is more difficult and is limited to a few plant species unless only simple distribution patterns following application to leaf surfaces are required. Direct measurement of the concentration of xenobiotics in the phloem sap of whole plants is only possible with some species (reviewed by Zimmermann and Milburn[18]), and useful measurements have also been made with phloem sap from certain excised plant parts.[19-22] However, it is always important to minimize the complicating effect of cuticular penetration following application to leaf surfaces. This may be accomplished by injecting solutions of chemicals into the plant; alternatively, compounds can be applied to leaf surfaces following abrasion or removal of the cuticle. In many respects the castor bean plant *(Ricinus communis)* is an excellent study system[23-25] since solutions of xenobiotics can be injected into the hollow petiole of the mature leaf and phloem sap can be collected directly from incisions in the stem.

Whatever the study system chosen, it is important to consider whether the conclusions drawn will apply to other plant species. Arguments for the similarity of transport processes in most plant species have been put forward.[26,27]

Many xenobiotics are readily metabolized in plants, and the translocation properties of the metabolites often are very different from those of the parent. It is, therefore, essential in any study system that the identity of the translocated compound is known. This is especially important when using radiolabeled materials since there is a temptation to use rapid but nonspecific techniques such as autoradiography or total counts from combustion to show distribution patterns. Techniques for studying the uptake and translocation of xenobiotics in plants have been reviewed in detail elsewhere.[26]

B. UPTAKE VIA ROOTS

1. Availability in Soil

A xenobiotic reaching soil is subject to the processes of sorption and loss which limit the amount of chemical available for uptake by plants[28] (see also Chapter 5). Degradation is the main loss process, although volatilization may be important for a few compounds that have high vapor pressure. Sorption of compounds is assessed as the soil-water distribution (K_d) and is generally proportional to the lipophilicity of the compounds and to the amount of soil organic matter. Sorption onto organic matter (K_{om}) can thus be estimated from K_d by the equation

$$K_d = K_{om} \cdot \frac{\% \text{ organic matter}}{100} \tag{4}$$

K_{om} can be estimated from K_{ow} by linear free energy relationships, such as that given by Briggs:[9]

$$\log K_{om} = 0.52 \log K_{ow} + 0.62 \tag{5}$$

Walker[29] showed that the uptake of atrazine into the shoots of wheat *(Triticum aestivum)* growing in 12 different soils was proportional to the estimated atrazine concentration in the soil water — i.e., uptake decreased with increasing amounts of organic matter in soil. In soils with organic carbon contents of 1 to 5%, lipophilic nonionized chemicals (i.e., log $K_{ow} > 4$ [see Section III.B]) will be strongly sorbed, with $K_d > 10$, while moderately lipophilic chemicals (i.e., log $K_{ow} = 2$ to 4) will be moderately sorbed, with $K_d = 1$ to 10. For polar and anionic compounds, the movement in soil water may carry the chemical toward or away from the plant roots, thereby affecting availability for uptake. Cations are more strongly bound to soil, by the process of ion exchange, giving K_d values of up to 50 for simple bases and of the order of 1000 for dications such as paraquat;[30] their availability for uptake is thus small. These factors have been reviewed recently with respect to herbicides, and the principles are broadly applicable.[31]

2. Uptake from Soil

Xenobiotics may be taken up by roots either via the vapor phase or via the water phase of the soil, the proportion taken up by each route depending largely on the physicochemical properties of the compound. The vapor/solution equilibrium of volatile organic compounds is described by Henry's law, and the Henry's law constant (H′) for a chemical is the ratio of saturated vapor concentration to water solubility:

$$H' = \frac{T}{273} \cdot \frac{v.p. \cdot 10^{-5}}{22.4 \; W.S.} \tag{6}$$

where v.p. is the vapor pressure (Pa) and W.S. is the water solubility (mol/l) measured at temperature T (K). Organic compounds with high Henry's law constants will thus have a high concentration in the vapor phase compared to the soil water and hence be taken up predominantly via the vapor phase. Those with low Henry's law constants will be taken up by roots from the aqueous phase, and compounds with constants of intermediate value will be taken up by both routes. This is shown diagrammatically in Figure 3 for a selection of possible soil contaminants.

Pesticides and industrial pollutants with a high Henry's law constant usually have a low affinity for water — i.e., they are relatively lipophilic. They are therefore unlikely to be translocated appreciably in plants, but may reach the foliage via the vapor phase (see Chapter 6). Thus, for example, herbicides such as trifluralin which are taken up via the vapor phase must exert their biological effect directly on the root system or on the emerging hypocotyl of the plant.

The efficiency of uptake by roots, by whichever route, can be expressed as the ratio of the concentration of chemical in the roots to the concentration of chemical in the surrounding medium. In the case of uptake via the aqueous

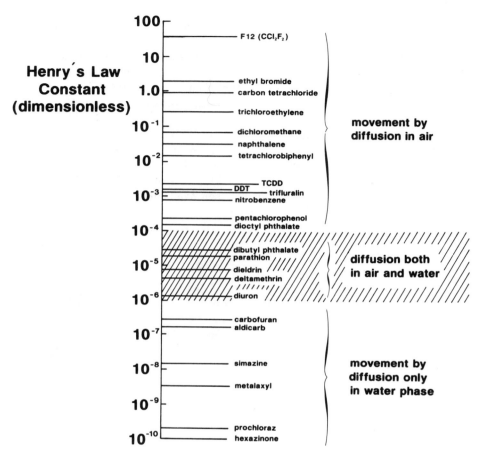

FIGURE 3. Pathways of movement of xenobiotics to plants through soil as determined by Henry's law constants.

phase, such as from nutrient solution, this ratio, called the root concentration factor (RCF),[32–34] is defined as

$$RCF = \frac{\text{concentration in roots}}{\text{concentration in external solution}} \tag{7}$$

The RCF is independent of concentration and independent of time once equilibrium has been established after a few hours.

a. Nonionized Compounds

The uptake of nonionized chemicals by roots from solution has been shown by Briggs et al.[34] to consist of two components. These are (1) an equilibration of the concentration in the aqueous phase inside the root with the concentration in the surrounding solution (contributing about 0.82 to the RCF) and (2) sorption of chemical onto the lipophilic root solids. The contribution of the latter component becomes important above a log K_{ow} of 1.5 and increases strongly as compounds become more lipophilic; separate measurements of sorption onto macerated root solids gave the relationship (expressed on a fresh weight basis)

$$\log RCF_{\text{macerated roots}} = 0.77 \log K_{ow} - 1.52 \qquad (8)$$

By measuring RCF values in barley *(Hordeum vulgare)* roots in intact plants for series of carbamoyloximes and phenylureas, Briggs et al.[34] (Figure 4) established the following relationship between RCF and lipophilicity, expressed as log K_{ow}:

$$\log (RCF - 0.82) = 0.77 \log K_{ow} - 1.52 \qquad (9)$$

This relationship also was found to hold for the uptake by a variety of plant species of a range of nonionized chemicals, measurements for which were taken from the literature.[34] Trapp and Pussemier,[35] studying the uptake into cut sections of bean *(Phaseolus vulgaris)* roots over 2 h, found that the RCF for 12 carbamates (spanning a log K_{ow} range of 1 to 3) was less than predicted from Equation 9, and Briggs and Evans (unpublished work) also observed the RCF of lipophilic compounds in beans to be only about one half of that observed in barley. A probable explanation for these observations is that bean roots contain less solids of a lipid-like nature than do barley roots.[35]

Attempts have been made to assign the uptake processes to three root compartments using analysis of the rates of uptake and/or efflux.[33,36] However, such experiments are difficult and easily confounded by metabolism of the compounds; furthermore, it has not been possible to identify what the putative compartments represent.

b. Weak Acids

The uptake of weak acids from aqueous solution by plant material increases as the pH of the surrounding solution decreases (Figure 5); this was first observed and explained for the uptake of 2,4-dichlorophenoxyacetic acid (2,4-D) by the alga *Chlorella pyrenoidosa*.[37] This phenomenon is an example of the ion trapping of weak acids (described in Section III.D), whereby the dissociated molecule is trapped in compartments of higher pH because anions permeate membranes less well than do the corresponding neutral molecules. In a series of substituted phenoxyacetic acids, Briggs et al.[13] estimated the permeability ratio of acid to anion (P_{AH}/P_{A}-) across a membrane to have a value between 180 and 4 x 10^5, the higher values of this ratio occurring for compounds of log K_{ow} ~2. Sorption onto root solids, except for weak acids of relatively high log K_{ow},

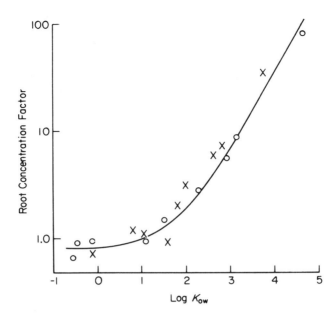

FIGURE 4. Relationship between the uptake of chemicals by plant roots (expressed as the root concentration factor) from nutrient solution at 24 h and their 1-octanol/water partition coefficients (as log K_{ow}): ○, O-methylcarbamoyloximes; x, substituted phenylureas. (From Briggs, G.G., Bromilow, R.H., and Evans, A.A. (1982), *Pestic. Sci.*, 13: 495. With permission.)

contributed little to the RCF compared to the large increase in concentration within the cells due to ion trapping. Thus, for weak acids, RCF is not only dependent on the pH of the bathing medium, but is also highest for acids of intermediate lipophilicity.

3. Translocation from Roots to Shoots

The efficiency of movement into the shoot from the roots, i.e., with the transpiration stream, is normally expressed as the transpiration stream concentration factor (TSCF),[32-34] which is defined as

$$TSCF = \frac{\text{concentration in xylem sap}}{\text{concentration in external solution}} \qquad (10)$$

Since the concentration in the xylem sap cannot be easily measured, it is usually estimated from measurements of the amount of compound accumulated for a known amount of transpiration (e.g., over a 24-h period). If the compound is appreciably metabolized over this period of time (t), the apparent TSCF so calculated must be corrected for the degradation rate (k) of the compound:[34]

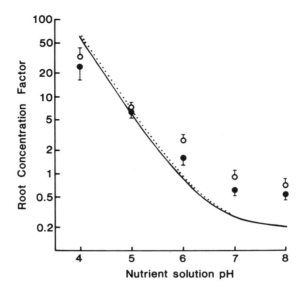

FIGURE 5. Uptake of substituted phenoxyacetic acids by barley roots (expressed as the root concentration factor [RCF]) at 24 h as a function of the pH of the nutrient solution: ○, 2,4-dichloro; ●, 3,5-dichloro. The vertical bars indicate the range of measured values. The lines are the RCF values predicted by the ion-trap mechanism; (----), 2,4-dichloro; (—), 3,5-dichloro. (From Briggs, G.G., Rigitano, R.L.O., and Bromilow, R.H. (1987), *Pestic. Sci.,* 19: 101. With permission.)

$$TSCF_{corrected} = (TSCF_{apparent} \cdot t \cdot k)/[1 - \exp(-kt)] \qquad (11)*$$

For passive uptake, the maximum value for the TSCF is, therefore, 1.0; i.e., the compound moves with the same efficiency as water.

a. Nonionized Compounds

Briggs et al.[34] measured the TSCF values for two series of nonionized compounds spanning a wide range of log K_{ow} values. The highest TSCF value was 0.9, which indicates passive movement of these chemicals into the xylem sap. The more polar compounds had a low value of TSCF, as did the very lipophilic ones, and the optimum TSCF value occurred with compounds whose log K_{ow} was around 1.8. Figure 6 shows their data, to which they fitted a Gaussian curve, the equation for which is

$$TSCF = 0.784 \exp -[(\log K_{ow} - 1.78)^2/2.44] \qquad (12)$$

The apparently anomalous point at log $K_{ow} = 1.57$ was 4-methylthiophenylurea, for which Briggs et al. measured in the shoots both parent compound and the

*This equation in the original paper contains an error introduced at the proof stage.

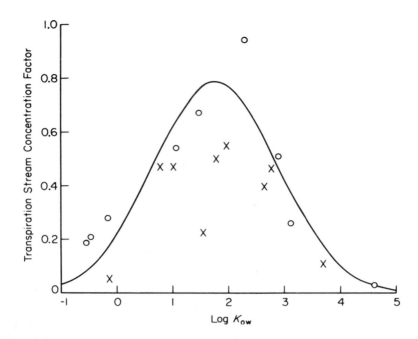

FIGURE 6. Relationship between the translocation of chemicals to barley shoots following uptake by roots over 24 h (expressed as the transpiration stream concentration factor) and their 1-octanol/water partition coefficients (as log K_{ow}): o, O-methylcarbamoyloximes; x, substituted phenylureas. (From Briggs, G.G., Bromilow, R.H., and Evans, A.A. (1982), *Pestic. Sci.,* 13: 445. With permission.)

oxidative sulfinyl and sulfonyl metabolites; these authors considered the low TSCF value of 0.22 to reflect in part rapid oxidation in the root to the sulfinyl derivative, which, having a log K_{ow} estimated at –0.2, would be expected to be poorly translocated. Literature values for the TSCF of a number of systemic pesticides were also found to fit the curve reasonably well, even though several plant species were involved. Hsu et al.[16] found a similar relationship in the translocation of cinmethylin and related compounds in detopped soybean plants using the pressure chamber technique, but their optimum TSCF occurred at the higher log K_{ow} value of 3.07:

$$TSCF = 0.7 \exp -[(\log K_{ow} - 3.07)^2/2.78] \qquad (13)$$

Compounds cannot move from root to shoot by a completely apoplastic pathway because the latter is blocked at the endodermis, between cortex and stele, by the Casparian strip. To reach the vascular system and thus the shoot, compounds must cross the plasmalemma, the membrane separating the apoplast and symplast, and TSCF is thus a measure of the ability of a compound to do this. The curve shown (Figure 6) therefore can be taken to represent membrane permeability plotted against lipophilicity, with maximum permeation occurring at a log K_{ow} value of approximately 1.8.

The reasons for this optimum log K_{ow} are not fully understood. Polar compounds will have difficulty crossing hydrophobic membranes, and indeed TSCF is reduced as log K_{ow} is decreased. However, the TSCF is also reduced above log K_{ow} ~2; this is not due to lipophilic compounds being removed from solution within the plant by partitioning onto lipid-like phases, for the values of RCF and TSCF do not change over 24 to 48 h of uptake. The equilibra for these processes are in fact reached within a few hours, after which time no more chemical is retained by roots and yet the TSCF is still maintained at a constant and often low value. It thus appears that, for reasons not certain, the more lipophilic compounds cross the endodermis much less efficiently than water, with compounds of log $K_{ow} > 4$ scarcely moving into the shoots at all.

Sorption to soil limits uptake of xenobiotics by plants (Section IV.B.1), and combining Equations 4, 5, and 12 enables predictions to be made of the influence of soil organic matter (Figure 7). Translocation via xylem is much reduced as the content of organic matter in soil is increased, with the uptake of lipophilic compounds being particularly affected. The log K_{ow} value for maximal translocation moves toward the more polar compounds as soil organic matter content, and, hence, sorption, is increased.

Certain compounds are so readily metabolized in plants that this process dominates their transport behavior. For example, Mc Farlane et al.[38] observed that the translocation of phenol and nitrobenzene from nutrient solution into the shoots of soybean was much less than predicted from Equation 12; however, these compounds were largely lost from the nutrient solution within 1 to 2 days, indicating that rapid metabolism in the roots was the probable cause of the very limited translocation.

However, in later experiments[15] utilizing eight plant species and measuring TSCF by loss of compound from solution and also by direct measurement in collected xylem sap, the mean TSCF for nitrobenzene was found to be 0.72 ± 0.07; this is very close to the value of 0.78 predicted from Equation 12 for nitrobenzene, for which log $K_{ow} = 1.85$. In these experiments, nitrobenzene was not only lost by metabolism in the plants, but was also appreciably lost as vapor from the transpiring leaves;[15,39] this interesting behavior, however, is unlikely to be typical of organic compounds in soil, for few except fumigants and industrial solvents have the volatility of nitrobenzene.

Although most of the physicochemical relationships given above have been developed using pesticides and related compounds, there is no reason to suppose that these relationships will not apply equally well to other compounds, such as industrial pollutants, provided that due consideration is given to complicating factors such as metabolism.

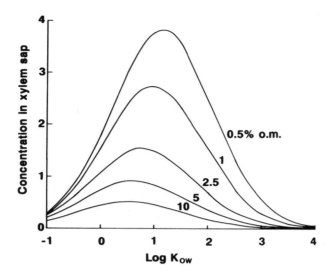

FIGURE 7. Influence of content of soil organic matter (o.m.) on the efficiency of uptake by plants of xenobiotics of differing lipophilicities. The contours represent concentrations in xylem sap for unit concentration of compound applied uniformly to soil containing 10% water by weight.

b. Weak Acids

The transport of weak acids from root to shoot is more complicated than that for nonionized compounds since acids are considerably ionized at the physiological pH of the cell sap of the cortex. As mentioned earlier (Section IV.B.2.b), the permeability of membranes to anions is very much lower than to neutral molecules, and so the movement of weak acids across the endodermis is not efficient. For example, the TSCF for 2,4-D in barley over 24 h is only 0.49 at a nutrient solution pH of 5.0.[13] This is despite the fact that the uptake by roots is high (RCF value of 7.10) due to ion trapping.

Since the uptake of weak acids by roots is dependent on the pH of the surrounding solution, it is not surprising that the TSCF values for such compounds are also pH dependent. Thus, the TSCF for 2,4-D from an external medium at pH 4.0 is 4.14 and at pH 7.0 is 0.04.[13] The values greater than unity for the TSCF indicate that the chemical is moving across the endodermis faster than water. This can be accomplished when the plant is expending energy in accumulating these weak acids in cell compartments, but this is not thought to indicate the presence of a specific carrier process.

4. Movement Within Shoots

Any compound which reaches the shoot and moves with the transpiration stream without loss will accumulate at sites of greatest transpiration, these being normally the mature leaves and, in many plants, the leaf edges. However, en route to the leaves in the xylem there is scope for reversible partitioning of nonionized chemicals onto the plant solids of the stem, and for weak acids there is also the possibility of ion trapping by compartments of high pH, including the adjacent phloem. For example, Uchida[40] showed that the distribution of dialkyl 1,3-dithiolan-2-ylidenemalonates between the upper and lower parts of the shoots of rice plants was determined by lipophilicity, the more lipophilic chemicals being more strongly retained by the stem bases. Briggs et al.[41] observed similar results with O-methylcarbamoyloximes and phenylureas in barley.

McCrady et al.[42] studied the breakthrough curves for 17 substituted benzenes perfused under pressure through 50-mm stem segments of soybean. The transport processes were also akin to column chromatography, with sorption of the chemicals to the tissue related to their K_{ow}; a mass transport model was derived, and this simulated well the movement and accumulation of the chemicals.

By analogy with the RCF, a stem concentration factor (SCF) has been defined[41] as

$$SCF = \frac{\text{concentration in stem}}{\text{concentration in external solution}} \qquad (14)$$

At any given time this will be given by the stem/xylem sap partition coefficient multiplied by that fraction of the external solution present in the xylem sap (i.e., the TSCF). For nonionized compounds, both of these entities are related solely to lipophilicity (expressed as log K_{ow}), and the same is therefore true of the SCF. For nonionized compounds on macerated stems, Briggs et al.[41] found that

$$\log SCF_{\text{macerated stems}} = 0.95 \log K_{ow} - 2.05 \qquad (15)$$

Barak et al.[43] found that the sorption of five systemic fungicides by macerated plant tissues was mainly due to partition onto lignin, and they found a relationship similar to Equation 15 for pepper, cotton, and bean plants.

Assuming that the contribution of the aqueous phase within cells to uptake by the stem was similar to that in the root (see Section IV.B.2), partitioning between stem and transpiration stream or xylem sap can be predicted from the equation

$$\log (K_{stem} - 0.82) = 0.95 \log K_{ow} - 2.05 \qquad (16)$$

Using this relationship together with that for the TSCF (Equation 12), Briggs et al.[41] predicted that the SCF, after equilibrium has been reached, would be given by

$$SCF = [10^{(0.95 \log K_{ow} - 2.05)} + 0.82] \, 0.784 \exp -[(\log K_{ow} - 1.78)^2/2.44] \quad (17)$$

This equation fit quite well the results obtained in barley for series of 0-methylcarbamoyloximes and substituted phenylureas, for which the SCF increased nearly linearly with log K_{ow} up to the most lipophilic compound tested (log K_{ow} = 4).

For weak acids, the SCF depends not only on the log K_{ow} of the undissociated acid and its pKa, but also on the permeability ratio (P_{AH}/P_{A}-) of the compound and on the pH of the xylem sap and the adjacent plant compartments.

C. UPTAKE AND TRANSPORT FOLLOWING FOLIAR APPLICATION

1. Uptake Through the Cuticle

The outer surfaces of plant leaves and of most fruits and stems are covered by a waxy layer, or cuticle, which serves as a barrier to water loss by the plant; this cuticle also slows the uptake of xenobiotics into plant foliage. These effects on the uptake of pesticides vary considerably between plant species, and can be reduced by the use of appropriate surfactants. In general, however, compounds of intermediate lipophilicity (log K_{ow} = 1 to 3) are taken up more easily than compounds outside that range, and weak acids (even with log K_{ow} in that range), such as many herbicides, are taken up relatively slowly. This is presumably due to the poor uptake of the dissociated compound. It is often overcome in the case of acidic herbicides by applying them as alkyl esters, which then undergo hydrolysis within the plant to the active acid. The role of the cuticle in the uptake of xenobiotics from deposits on the leaf surface and in uptake from the vapor phase is dealt with in the chapters in Part Three.

2. Transport in Xylem

Passage of a chemical across the plant cuticle may limit the amount available for transport within the plant; however, this aside, xylem transport of xenobiotics following uptake from leaves obeys the same general relationships as following uptake by roots. However, compounds taken up by roots must cross into the symplasm at the endodermis in order to move to the shoots in the xylem (Section IV.B.3), and this reduces the xylem translocation of the more polar compounds, which do not cross the plasmalemma very efficiently. Compounds applied to the shoots do not have to cross this membrane to enter the xylem, thus allowing the more efficient movement of polar compounds, as well as those of intermediate lipophilicity, in the transpiration stream. The movement of lipophilic compounds is limited by their sorption onto plant solids, and for foliar application this includes the cuticle itself, which strongly sorbs compounds of relatively high log K_{ow}.

3. Transport in Phloem

The mechanism of transport of both endogenous compounds and xenobiotics in phloem has long been a subject of discussion. Specific carrier processes have been shown to load sucrose (and related transport sugars) into the phloem, and the difference in osmotic pressure so created draws water into the sieve tubes and, hence, displaces the contents toward the zones of sugar utilization; i.e., mass flow occurs toward the "sinks". Roles for specific carriers have been postulated from time to time for other endogenous compounds such as amino acids and plant growth hormones, but the evidence is inconclusive. Specific carriers do not appear to recognize xenobiotics; the rest of this section is confined to consideration of the transport of these compounds in phloem.

As discussed earlier, any compound which is translocated from the roots to the shoots must cross the plasmalemma from the apoplast to the symplast at the endodermis. There is no reason to believe that the membrane separating phloem and xylem vessels is very different from that in the endodermis, and it follows that compounds mobile in xylem following root uptake must be capable of entering the phloem. From these considerations, Tyree et al.[44] postulated that, while many compounds are capable of entering phloem, only compounds that are retained in the phloem sieve tubes are well translocated; those which move freely across the separating membrane show an overall pattern of xylem translocation because of the much greater volume of flow in the xylem vessels compared to the phloem (50- to 100-fold). These ideas are now widely accepted and form the basis of the "intermediate permeability theory".

Phloem translocation of a compound is dependent on two processes. The first is uptake of the compound by the phloem sieve tubes from the surrounding tissues, and the second process is the retention of the compound in the phloem during translocation. Thus, nonionized chemicals of intermediate permeability such as oxamyl and aldoxycarb, with log K_{ow} values of –0.47 and –0.57, respectively, show some degree of phloem movement[25] because they can enter the phloem in areas of high concentration close to the point of application but are too polar to move out of the phloem readily. Very polar nonionized compounds do not cross the membranes sufficiently to be able to enter the phloem vessels, and so they are not translocated. More lipophilic compounds (log K_{ow} = 1 to 3), which cross membranes very readily, can enter the phloem easily but immediately diffuse back into the greater volume of the xylem.

When it was realized that the majority of phloem-mobile xenobiotics were weak acids, it was assumed that they were loaded into the phloem by a carrier mechanism that recognized the acid group.[45,46] It is now realized that such compounds are phloem mobile partly because of ion-trapping (Section III.D) in the more basic phloem (pH ~8) relative to the apoplast (pH ~5.5), including the xylem. The phloem translocation of weak acids is now seen simply as an extension of the intermediate permeability theory; this can explain the modest phloem transport of stronger acids, for which accumulation by ion trapping is negligible.

The optimum physicochemical properties of a weak acid for diffusion into the phloem are for it to be undissociated as much as possible (i.e., high pKa) and for the undissociated molecule to have a log K_{ow} ~1.8 (Section IV.B.3). On the other hand, to be retained effectively in phloem and, therefore, show long-distance transport, an acid requires a high degree of ionization (i.e., low pKa), and the small proportion of undissociated molecules present need to be very polar (log K_{ow} < 1). Thus, the optimum values for efficient phloem transloca-tion of weak acids must be a compromise between these conflicting require-ments for the two contributing processes.

From studies on the phloem translocation of many compounds using the castor bean system (Section IV.A), it has been shown that one of the com-pounds translated most efficiently over long distances is maleic hydrazide.[47] This compound has a pKa value of 5.65, which is near the optimum for accumulation by the phloem but which is too high for good retention during translocation; its log K_{ow}, however, is –0.63, which is very good for retention but too low for efficient uptake. Since such compounds are translocated so effectively in the phloem, it must be assumed that there is considerable overlap between the ranges of optimum physicochemical properties for the two pro-cesses.[12] Rigitano et al.[47] compared measured values of phloem loading for weak acids in the leaves of castor beans with values predicted on the basis of ion trapping (Equation 3) using values of the permeability ratio P_{AH}/P_{A^-} obtained from barley roots[13] (Figure 8). The concentration of maleic hydrazide was 10.4 times greater in the phloem sap than in the bulk leaf tissue (averaging over vacuoles and cytoplasm) and was accurately predicted, although predictions for a series of substituted phenoxyacetic acids were two to three times higher than the measured values. Nonetheless, agreement was sufficiently good to encourage the authors to believe that simple physicochemical processes were the basis for the accumulation and transport of weak acids in phloem.

The herbicide glyphosate is also translocated extremely well in phloem, and foliar applications give good control of rhizomatous and deep-rooted weeds.[48] The phloem transport of this polar compound (log K_{ow} < 0) is more difficult to understand quantitatively since it contains three separate acid functions (pKa values of < 2, 2.3, and 5.6) and a basic amino group (pKa = 10.2). The possible ionizations are thus very complex, and at present very little is known about the transport behavior of positively charged species and of zwitterions. Organic cations are attracted by the negative charge of cells, although, based on what is known for anions (Section III.D), from permeability considerations it might be expected that nonquaternized cations would accu-mulate in compartments of low pH. Such compartments include the apoplast and cell vacuoles, both of which have pH ~5.5. Such effects would reduce phloem translocation since accumulation in the vacuoles of normal cells would prevent the compounds from reaching the specialized phloem cells, which do

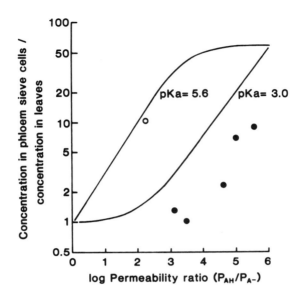

FIGURE 8. Accumulation of weak acids in phloem sieve cells in leaves of castor bean. The curves show the predicted accumulation for weak acids having different pKa values and for cell membranes having different ratios of permeabilities to the undissociated and dissociated forms of the acids. Points show the observed accumulation: ●, substituted phenoxyacetic acids, pKa ~3.0; ○, maleic hydrazide, pKa = 5.65. (From Rigitano, R.L.O., Bromilow, R.H., Briggs, G.G., and Chamberlain, K. (1987), *Pestic. Sci.*, 19: 113. With permission.)

not contain vacuoles, and any reaching the vascular tissues would accumulate in the more acidic medium of the xylem.

These ideas, based largely on the work of the authors using the castor bean system, are in accord with the model for phloem translocation described by Kleier,[49] which shows how the ion-trapping effect involved in the transport of weak acids fits into the intermediate permeability theory first proposed by Tyree et al.[44] The model of Kleier has been tested and found to hold for series of benzoic acids and pyridinium cations in pinto beans.[19]

D. DISTRIBUTION PATTERNS WITHIN PLANTS AND THEIR RELATION TO PHYSICOCHEMICAL PROPERTIES

The factors affecting the movement of organic chemicals in plants have been discussed above, and consideration is now given to the consequences for the overall distribution of xenobiotics in plants. This distribution is normally measured by autoradiography following application of radiolabeled compounds. Autoradiography shows in great detail where the radioactivity accumulates, but has the disadvantage that it shows only the position of the radioisotope and does not identify the compound. If the compound under study is unlikely to be stable over the course of the experiment, then the nature of the transported radioactivity must be identified. Only then is it possible to relate the transport of parent compound and its metabolites to their respective physicochemical properties.

The relationships between transport in plants and chemical structure have been under consideration for many years, particularly with regard to pesticides; the reader is referred to reviews by Crafts and Crisp,[50] Crisp,[45] Christ,[51] Price,[52] Edgington,[53] Jacob and Neumann,[54] Shephard,[55] Lichtner,[21,56] Giaquinta,[57] Bromilow and Chamberlain,[12,27] and Bromilow et al.[58,59] Much of this early work can now be understood and fitted into our present understanding of the transport of xenobiotics. The only appreciable point of difference is that the role once postulated for specific carrier mechanisms, particularly with regard to phloem transport, now seems unlikely, at least for the compounds currently under consideration.

Nonionized chemicals entering plants, whether by uptake via roots or via foliage, move predominantly with the water flow in the xylem. This movement is subject to a process analogous to column chromatography (Section IV.B.3), and the transport of lipophilic compounds is substantially slowed by reversible partitioning onto the lipid-like materials of the plant solids. Transport via phloem occurs only for rather polar nonionized compounds, and even for these the amounts accumulating in phloem sinks are very small.[25]

Over periods of several days, nonionized compounds moving in the xylem tend to accumulate in the interveinal spaces and at the leaf margins. In dicotyledonous plants, compounds applied to a point on a leaf will give rise to a wedge-shaped distribution pattern (the "apoplastic wedge"); in monocotyledonous plants, accumulation occurs at the leaf tips, often to high concentrations.

Although much of this work has been done with pesticides, it appears equally applicable to other xenobiotics. For example, Harvey et al.[60] studied the uptake of the explosive RDX (hexahydro-1,3,5-trinitro-1,3,5-triazine) from hydroponic solution by bean plants. Concentrations of parent RDX in roots and stem remained essentially constant after 1 and 7 days of uptake, over which period concentrations in the leaves increased fivefold; the amount of uptake was consistent with that expected for such a compound with log K_{ow} = 0.87.

Topp et al.[61] claimed that molecular weight was a more suitable parameter by which to predict uptake of chemicals into barley plants (roots and foliage considered together) from soil. However, their range of chemicals appears to show a strong correlation between molecular size and K_{ow} and lacks useful compounds of log K_{ow} < 2.6. Travis and Arms[62] listed values taken from the literature of bioconcentration factors into plant foliage for 29 chemicals spanning log K_{ow} values from 1 to 10 and found the bioconcentration factor to be inversely proportional to the square root of K_{ow}. However, in all such studies there are several processes contributing to uptake (especially where all the plant parts are considered together), and it seems unlikely that simple regressions can reliably predict behavior for compounds spanning a wide range of physicochemical properties.

Turning to compounds more mobile in phloem, autoradiography or other analysis indicates the efficiency with which compounds applied to or reaching, for example, mature leaves are transported to phloem sinks such as roots or new foliar growth. As discussed above (Section IV.C.3), most compounds are able to enter the phloem vessels, but movement of many compounds in phloem is limited by their crossing over to the xylem and returning to the transpiring leaves. Weak acids are the main class of compounds moving in phloem, and these also move freely in xylem. However, their movement in both these vascular tissues is also limited by a process akin to chromatography in which these compounds are held by cells against the moving phase (be it phloem or xylem) not by partitioning, as occurs with nonionized compounds, but apparently by ion trapping of such weak acids. This process influences both long-distance transport in plants and distribution within leaves. With very lipophilic acids (log $K_{ow} > 4$), both the processes of partitioning and ion trapping may limit movement.

Distributions resulting from these processes are illustrated in Figure 9, which shows autoradiographs of mature and apical leaves of castor bean plants after 24 h of uptake by roots of ^{14}C-labeled compounds.[63] The nonionized 4-chlorophenylurea (log $K_{ow} = 1.8$) moves only to the mature leaves, the traces of ^{14}C in the apical leaf being due to metabolites; within each mature leaf the compound is concentrated in the interveinal spaces and leaf margins, from which the transpiration losses are greatest. In contrast, 4-fluorophenoxyacetic acid (log $K_{ow} = 1.64$, pKa = 3.1) appears at high concentration in the phloem sink of the apical leaf, and within the mature leaves it is concentrated around the veins. Such strong accumulation by ion trapping is also in accord with its very limited movement to roots following application to leaves;[47] much of the chemical accumulates in the stem directly below the exporting mature leaves. Not all weak acids are strongly held by tissues in this way, for the sensitivity of the ion-trapping mechanism varies according to the pKa of the acid (Equation 3). In general, it seems that the more polar and/or strongly acidic compounds are those best able to move over long distances via phloem (e.g., glyphosate, maleic hydrazide, imazapyr, clopyralid).

Phytotoxicity can limit the movement of xenobiotics — particularly, of course, that of herbicides, since these are designed to damage and kill all or certain plant species. Many herbicides cause membrane disruption, usually as a secondary effect, and this destroys the integrity of the transport systems, particularly phloem. If such effects occur close to the point of application of a herbicide, they will interfere with subsequent translocation to other parts of the plant. Rapid phytotoxicity — e.g., that caused by dinitrophenols, paraquat, glufosinate, etc. — is easily observed, and it is clear that the uptake and transport behavior of such compounds cannot be successfully modeled under these circumstances. Carter et al.[64] studied the systemic activity in leaves of

FIGURE 9. Autoradiographs showing the distribution within the leaf blades (mature leaf(s) and apical leaf presented) of castor bean 24 h after application to roots via nutrient solution: (a) 4-chlorophenylurea; (b) 4-fluorophenoxyacetic acid. (From Rigitano, R.L.O. (1985), Ph.D. thesis, University of London, London, England. With permission.)

cucumber *(Cucumis sativus)* of a range of antifungal trichloroethylformamides applied to the roots in sand culture. Accumulation of some of these compounds in leaves gave rise to phytotoxic symptoms; a compound of intermediate lipophilicity (log K_{ow} ~2) damaged the leaves by accumulating in the interveinal spaces (Figure 10a), whereas a more lipophilic compound damaged the veinal areas (Figure 10b), presumably because sorption to tissue limited its further movement.

However, more subtle effects on transport systems may occur that do not immediately lead to visible symptoms of damage. Mc Farlane and Pfleeger[65] have developed a system in which the transpiration of plants grown in nutrient solution in a chamber is monitored continuously; thus, any phytotoxic effects

a   b

FIGURE 10. Phytotoxicity to cucumber leaves caused by different analogues of trichloroethylformamide fungicides applied to roots in sand culture. See text for details. (See Reference 64.)

caused by the addition of a xenobiotic to the nutrient solution can be rapidly assessed. Phloem transport has been observed to be inhibited quite rapidly by the herbicides glyphosate and chlorsulfuron, although visible signs of injury take days to appear. Geiger and co-workers[66-68] found that foliar application of glyphosate rapidly inhibited the allocation of carbon to starch during photosynthesis in sugar beet *(Beta vulgaris)*, although sucrose levels and export were maintained. As a consequence of this reduction in starch levels during the light period, sucrose export during the subsequent dark period was inhibited, and since glyphosate is distributed via bulk flow in the phloem, its movement also was reduced. Likewise, chlorsulfuron has been shown to affect its own translocation by inhibiting assimilate translocation from the treated leaves of common pennycress *(Thlaspi arvense)* and Tartary buckwheat *(Fagopyrum tataricum)*.[69]

It is now appropriate to reconsider the question of the degree to which transport in a test plant species is indicative of the behavior in other plant species. Chamberlain et al.[70] compared the distribution of six substituted phenoxyacetic acids in two plant species. Application to castor bean was by injection into the petioles of mature leaves and to barley was by application (with surfactant) to a central section of the leaf blade. After 24 h, the proportion of each chemical moved in the phloem (Figure 11) was greatest for the compounds of intermediate lipophilicity and was very similar for the two plant species. Taken together with the previously discussed results on xylem transport (Section IV.B.3.a), these results indicate that measurements of transport in convenient experimental systems such as castor bean are applicable to crop species.

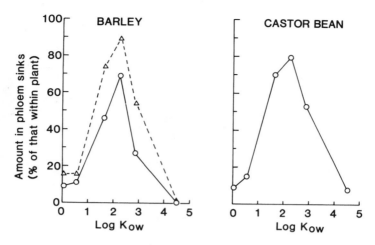

FIGURE 11. Phloem translocation of substituted phenoxyacetic acids in barley and castor bean: ○, 24 h after application; △, 48 h after application. (From Chamberlain, K., Briggs, G.G., Bromilow, R.H., Evans, A.A., and Fang, C.Q. (1987), *Aspects Appl. Biol.,* 14: 293. With permission.)

Distribution patterns and autoradiographs have been published illustrating the transport of many xenobiotics, particularly pesticides, in plants. The examples of compounds discussed above exemplify virtually all the observed patterns of behavior, with no major differences being noted between different plant species except where selective metabolism is occurring.

V. CONCLUSIONS

The processes of uptake and translocation of organic molecules in plants appear to be largely determined by the physicochemical properties of the compounds. Bromilow and Chamberlain[12] have produced a simple diagram (Figure 12) indicating the properties required by nonionized compounds and weak acids for the various types of systemic transport. It should be noted that the divisions between the categories are not quite so clear-cut in practice, but the diagram does indicate the main characteristics of behavior. For example, nonionized compounds of log $K_{ow} > 4$ applied to foliage may move slightly in xylem given sufficient time, although pesticides applied to soil usually fail to have action against pests or diseases on leaves if they have log $K_{ow} > 3$. The possibility of such compounds moving through the vapor phase also should be considered.

One of the difficulties in choosing relationships to use in models lies in assessing their general applicability. Many studies have included only compounds with a restricted range of properties; extrapolation is generally unwise,

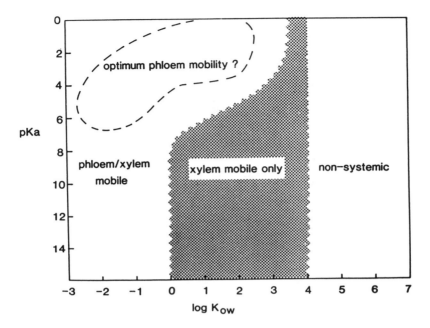

FIGURE 12. Physicochemical properties required of weak acids and nonionized compounds for various types of systemic behavior. (From Bromilow, R.H. and Chamberlain, K., (1989), in *Mechanisms and Regulation of Transport Processes,* R.K. Atkin and D.R. Clifford, Eds., Monogr. 18, British Plant Growth Regulator Group, p. 113. With permission.)

especially if the postulated relationships are derived from regression analyses without a full understanding of the processes involved.

Several attempts have been made, based on the concepts and approaches discussed in this chapter, to predict and model the uptake and transport of chemicals in plants. Topp et al.,[61] Ryan et al.,[71] Schramm and Hutzinger,[72] and Trapp et al.[73] put forward methods to assess the uptake from soil of nonionized chemicals based on their physicochemical properties and persistence. Robinson and Dunham[74] modeled the uptake of chlorotriazines from soil, and more complex network models have been developed by Boersma et al.,[75] McCoy,[76] Lindstrom et al.,[77] and Boersma et al.[78] Kleier[49] has produced a computer model that predicts phloem transport of simple xenobiotics from log K_{ow} and pKa values. All of these models gave reasonable predictions for laboratory experiments, but extrapolating such estimates to field situations requires due consideration of complicating factors such as availability in soil, penetration across leaf cuticles, and metabolism. However, bearing these in mind, from our present knowledge of the uptake and transport of xenobiotics in plants it should be possible to predict approximately the behavior of pesticides or industrial pollutants, and the following chapters address the modeling of these processes.

REFERENCES

1. Worthing, C.E. and Hance, R.J. (1991), *The Pesticide Manual,* 9th ed., The British Crop Protection Council, London.
2. Ashton, F.M. and Crafts, A.S. (1981), *Mode of Action of Herbicides,* John Wiley & Sons, New York.
3. Howard, P.H., Ed. (1991), *Handbook of Environmental Fate and Exposure Data for Organic Chemicals,* Vol. 3, Lewis Publishers, Boca Raton, FL.
4. Howard, P.H., Ed. (1991), *Handbook of Environmental Fate and Exposure Data for Organic Chemicals,* Vols. 1 & 2, Lewis Publishers, Boca Raton, FL.
5. Hutzinger, O., Ed. (1991), *The Handbook of Environmental Chemistry,* Vol. 3, Anthropogenic Compounds, Springer-Verlag, Berlin.
6. Fujita, T., Iwasa, J., and Hansch, C.H. (1964), A new substituent constant, π, derived from partition coefficients, *J. Am. Chem. Soc.,* 86: 5175.
7. Dunn III, W.J., Block, J.H., and Pearlman, R.S., Eds. (1986), *Partition Coefficient, Determination and Estimation,* Pergamon Press, New York.
8. Hansch, C.H. and Leo, A. (1979), *Substituent Constants for Correlation Analysis in Chemistry and Biology,* John Wiley & Sons, New York.
9. Briggs, G.G. (1981), Theoretical and experimental relationships between soil adsorption, octanol-water partition coefficients, water solubility, bioconcentration factors, and the parachor, *J. Agric. Food Chem.,* 29: 1050.
10. Albert, A. and Serjeant, E.P. (1962), *Ionization Constants of Acids and Bases. A Laboratory Manual,* Methuen and Co., London.
11. Perrin, D.D., Dempsey, B., and Serjeant, E.P. (1981), *pKa Prediction for Organic Acids and Bases,* Chapman and Hall, London.
12. Bromilow, R.H. and Chamberlain, K. (1989), Designing molecules for systemicity, in *Mechanisms and Regulation of Transport Processes,* R.K. Atkin and D.R. Clifford, Eds., Monograph 18, British Plant Growth Regulator Group, p. 113.
13. Briggs, G.G., Rigitano, R.L.O., and Bromilow, R.H. (1987), Physicochemical factors affecting uptake by roots and translocation to shoots of weak acids in barley, *Pestic. Sci.,* 19: 101.
14. Raven, J.A. (1975), Transport of indolacetic acid in plant cells in relation to pH and electrical potential gradients, and its significance for polar IAA transport, *New Phytol.,* 74: 163.
15. Mc Farlane, C., Pfleeger, T., and Fletcher, J. (1990), Effect, uptake and disposition of nitrobenzene in several terrestrial plants, *Environ. Toxicol. Chem.,* 9: 513.
16. Hsu, F.C., Marxmiller, R.L., and Yang, A.Y.S. (1991), Study of root uptake and xylem translocation of cinmethylin and related compounds in detopped soybean roots using a pressure chamber technique, *Plant Physiol.,* 93: 1573.
17. Schroll, R. and Scheunert, I. (1992), A laboratory system to determine separately the uptake of organic chemicals from soil by plant roots and by leaves after vaporization, *Chemosphere,* 24: 97.
18. Zimmermann, M.H. and Milburn, J.A., Eds. (1975), *Encylopedia of Plant Physiology,* New Series, Vol. 1, Transport in Plants. I. Phloem Transport, Springer-Verlag, Berlin.

19. Hsu, F.C., Kleier, D.A., and Melander, W.R. (1988), Phloem mobility of xenobiotics. II. Bioassay testing of the unified mathematical model, *Plant Physiol.,* 86: 811.
20. Groussol, J., Delrot, S., Caruhel, P., and Bonnemain, J-L. (1986), Design of an improved exudation method for phloem sap collection and its use for the study of phloem mobility of pesticides, *Physiol. Vég.,* 24: 123.
21. Lichtner, F.T. (1986), Phloem transport of agricultural chemicals, in *Plant Biology,* Vol. 1, Phloem Transport, J. Cronshaw, W.J. Lucas, and R.T. Giaquinta, Eds., Alan R. Liss, New York, p. 601.
22. Neumann, S., Grimm, E., and Jacob, E. (1985), Transport of xenobiotics in higher plants. I. Structural prerequisites for translocation in the phloem, *Biochem. Physiol. Pflanzen,* 180: 257.
23. Milburn, J.A. (1972), Phloem transport in *Ricinus, Pestic. Sci.,* 3: 653.
24. Chamberlain, K., Burrell, M.M., Butcher, D.N., and White, J.C. (1984), Phloem transport of xenobiotics in *Ricinus communis* var. Gibsonii, *Pestic. Sci.,* 15: 1.
25. Bromilow, R.H., Rigitano, R.L.O., Briggs, G.G., and Chamberlain, K. (1987), Phloem translocation of non-ionised chemicals in *Ricinus communis, Pestic. Sci.,* 19: 85.
26. Bromilow, R.H., Chamberlain, K., and Briggs, G.G. (1986), Techniques for studying uptake and translocation of pesticides in plants, *Aspects Appl. Biol.,* 11: 29.
27. Bromilow, R.H. and Chamberlain, K. (1991), Pathways and mechanisms of transport of herbicides in plants, in *Target Sites for Herbicide Action,* R.C. Kirkwood, Ed., Plenum Press, New York, p. 245.
28. Briggs, G.G. (1984), Factors affecting the uptake of soil-applied chemicals by plants and other organisms, in *Proc. BCPC Symp. Soils and Crop Protection Chemicals,* R. Hance, Ed., Monograph 27, British Crop Protection Council, Croydon, p. 35.
29. Walker, A. (1972), Availability of atrazine to plants in different soils, *Pestic. Sci.,* 3: 139.
30. Nicholls, P.H. and Evans, A.A. (1991), Sorption of ionisable organic compounds by field soils. II. Cations, bases and zwitterions, *Pestic. Sci.,* 33: 331.
31. Leake, C.R. (1991), Fate of soil-applied herbicides: factors influencing delivery of active ingredients to target sites, in *Target Sites for Herbicide Action,* R.C. Kirkwood, Ed., Plenum Press, New York, p. 189.
32. Shone, M.G.T. and Wood, A.V. (1974), A comparison of the uptake and translocation of some organic herbicides and a systemic fungicide by barley. I. Absorption in relation to physico-chemical properties, *J. Exp. Bot.,* 25: 390.
33. Shone, M.G.T., Bartlett, B.O., and Wood, A.V. (1974), A comparison of the uptake and translocation of some organic herbicides and a systemic fungicide by barley. I. Relationship between uptake by roots and translocation to shoots, *J. Exp. Bot.,* 25: 401.
34. Briggs, G.G., Bromilow, R.H., and Evans, A.A. (1982), Relationships between lipophilicity and root uptake and translocation of non-ionised chemicals by barley, *Pestic. Sci.,* 13: 495.
35. Trapp, S. and Pussemier, L. (1991), Model calculations and measurements of uptake and translocation of carbamates by bean plants, *Chemosphere,* 22: 327.

36. Balke, N.E. and Price, T.P. (1988), Relationship of lipophilicity to influx and efflux of triazine herbicides in oat roots, *Pestic. Biochem. Physiol.*, 30: 228.
37. Wedding, R.T. and Erickson, L.C. (1957), The role of pH in the permeability of *Chlorella* to 2,4-D, *Plant Physiol.*, 32: 503.
38. Mc Farlane, J.C., Pfleeger, T., and Fletcher, J. (1987), Transpiration effect on the uptake and distribution of bromacil, nitrobenzene, and phenol in soybean plants, *J. Environ. Qual.*, 16: 372.
39. Fletcher, J.S., Mc Farlane, J.C., Pfleeger, T., and Wickliff, C. (1990), Influence of root exposure concentration on the fate of nitrobenzene in soybean, *Chemosphere*, 20: 513.
40. Uchida, M. (1980), Affinity and mobility of fungicidal dialkyl dithiolanylidenemalonates in rice plants, *Pestic. Biochem. Physiol.*, 14: 249.
41. Briggs, G.G., Bromilow, R.H., Evans, A.A., and Williams, M.R. (1983), Relationships between lipophilicity and the distribution of nonionised chemicals in barley shoots following uptake by the roots, *Pestic. Sci.*, 14: 492.
42. McCrady, J.K., Mc Farlane, C., and Lindstrom, F.T. (1987), The transport and affinity of substituted benzenes in soybean stems, *J. Exp. Bot.*, 38: 1875.
43. Barak, E., Dinoor, A., and Jacoby, B. (1983), Adsorption of systemic fungicides and a herbicide by some components of plant tissues in relation to some physicochemical properties of the pesticides, *Pestic. Sci.*, 14: 213.
44. Tyree, M.T., Peterson, C.A., and Edgington, L.V. (1979), A simple theory regarding ambimobility of xenobiotics with special reference to the nematicide oxamyl, *Plant Physiol.*, 63: 367.
45. Crisp, C.E. (1972), The molecular design of systemic insecticides and organic functional groups in translocation, in *Pesticide Chemistry, Proc. 2nd IUPAC Congress*, A.S. Tahori, Ed., Gordon and Breach, New York, p. 211.
46. Crisp, C.E. and Look, M. (1979), Phloem loading and transport of weak acids, in *Advances in Pesticide Science*, Geissbühler, H., Brooks, G.T. and Kearney, P.C., Eds., Pergamon Press, Oxford, p. 430.
47. Rigitano, R.L.O., Bromilow, R.H., Briggs, G.G., and Chamberlain, K. (1987), Phloem translocation of weak acids in *Ricinus communis, Pestic. Sci.*, 19: 113.
48. Grossbard, E. and Atkinson, D., Eds. (1985), *The Herbicide Glyphosate*, Butterworths, London.
49. Kleier, D.A. (1988), Phloem mobility of xenobiotics I. Mathematical model unifying the weak acid and intermediate permeability theories, *Plant Physiol.*, 86: 803.
50. Crafts, A.S. and Crisp, C.E. (1971), *Phloem Transport in Plants*, W.H. Freeman, San Francisco.
51. Christ, R.A. (1979), Physiological and physicochemical requisites for the transport of xenobiotics in plants, in *Advances in Pesticide Science: Proc. 4th IUPAC Congress*, Vol. 3, H. Geissbühler, Ed., Pergamon Press, Oxford, p. 420.
52. Price, C.E. (1979), Movement of xenobiotics in plants — perspectives, in *Advances in Pesticide Science: Proc. 4th IUPAC Congress* Vol. 3, H. Geissbühler, Ed., Pergamon Press, Oxford, p. 401.
53. Edgington, L.V. (1981), Structural requirements of systemic fungicides, *Annu. Rev. Phytopathol.*, 19: 107.

54. Jacob, F. and Neumann, S. (1983), Quantitative determination of mobility of xenobiotics and criteria of their phloem and xylem mobility, in *Pesticide Chemistry: Human Welfare and the Environment,* Vol. 1, J. Miyamoto and P.C. Kearney, Eds., Pergamon Press, Oxford, p. 357.

55. Shephard, M.C. (1985), Fungicide behaviour in the plant: systemicity, in *Fungicides for Crop Protection,* I.M. Smith, Ed., Monograph 31, British Crop Protection Council, London, p. 99.

56. Lichtner, F.T. (1984), Phloem transport of xenobiotic chemicals, *Whats New Plant Physiol.,* 15: 29.

57. Giaquinta, R.T. (1985), Physiological basis of phloem transport of agrichemicals, in *Bioregulators for Pest Control,* P.A. Hedin, Ed., American Chemical Society, Washington, D.C., p. 7.

58. Bromilow, R.H., Chamberlain, K., and Evans, A.A. (1991), Molecular structure and properties of xenobiotics in relation to phloem translocation, in *Recent Advances in Phloem Transport and Assimilate Compartmentation,* Bonnemain, J.L., Delrot, S., Lucas, W.J., and Dainty, J., Eds., Ouest Editions, Nantes, France, p. 332.

59. Bromilow, R.H., Chamberlain, K., and Evans, A.A. (1990), Physicochemical aspects of phloem translocation of herbicides, *Weed Sci.,* 38: 305.

60. Harvey, S.D., Fellows, R.J., Cataldo, D.A., and Bean, R.M. (1991), Fate of the explosive hexahydro-1,3,5-trinitro-1,3,5-triazine (RDX) in soil and bioaccumulation in bush bean hydroponic plants, *Environ. Toxicol. Chem.,* 10: 845.

61. Topp, E., Scheunert, I., Attar, A., and Korte, F. (1986), Factors affecting the uptake of ^{14}C-labeled organic chemicals by plants from soil, *Ecotoxicol. Environ. Saf.,* 1: 365.

62. Travis, C.C. and Arms, A.D. (1988), Bioconcentration of organics in beef, milk and vegetation, *Environ. Sci. Technol.,* 22: 271.

63. Rigitano, R.L.O. (1985), Physicochemical Factors Affecting Translocation and Distribution of Xenobiotics in Plants, Ph.D. thesis, University of London. London, England.

64. Carter, G.A., Chamberlain, K., and Wain, R.L. (1975), Investigations on fungicides. XVII. The systemic antimildew activity of alkoxy-, alkylamino- and alkylthio-trichloroethylformamides, *Ann. Appl. Biol.,* 79: 313.

65. Mc Farlane, J.C. and Pfleeger, T. (1987), Plant exposure chambers for study of toxic chemical-plant interactions, *J. Environ. Qual.,* 16: 361.

66. Gougler, J.A. and Geiger, D.R. (1984), Carbon partitioning and herbicide transport in glyphosate-treated sugar beet *(Beta vulgaris), Weed Sci.,* 32: 546.

67. Geiger, D.R., Kapitan, S.W., and Tucci, M.A. (1986), Glyphosate inhibits photosynthesis and allocation of carbon to starch in sugar beet leaves, *Plant Physiol,* 82: 468.

68. Servaites, J.C., Tucci, M.A., and Geiger, D.R. (1987), Glyphosate effects on carbon assimilation, ribulose bisphosphate carboxylase activity and metabolite levels in sugar beet leaves, *Plant Physiol,* 85: 370.

69. Vanden Born, W.H., Bestman, H.D., and Devine, M.D. (1988), The inhibition of assimilate translocation by chlorsulfuron as a component of its mechanism of action, in *Factors Affecting Herbicidal Activity and Selectivity,* Proc. EWRS Symp., p. 69.

70. Chamberlain, K., Briggs, G.G., Bromilow, R.H., Evans, A.A., and Fang, C.Q. (1987), The influence of physicochemical properties of pesticides on uptake and translocation following foliar application, *Aspects Appl. Biol.,* 14: 293.
71. Ryan, J.A., Bell, R.M., Davidson, J.M., and O'Connor, G.A. (1988), Plant uptake of non-ionic organic chemicals from soils, *Chemosphere*, 17: 2299.
72. Schramm, K.W. and Hutzinger, O. (1990), UNITTree: A model to estimate the fate of lipophilic compounds in plants, *Toxicol. Environ. Chem.,* 26: 61.
73. Trapp, S., Matthies, M., Scheunert, I., and Topp, E.M. (1992), Modeling the bioconcentration of organic chemicals in plants, *Environ. Sci. Technol.,* 24: 1246.
74. Robinson, R.C. and Dunham, R.J. (1982), The uptake of soil-applied chlorotriazines by seedlings and its prediction, *Weed Res.,* 22: 223.
75. Boersma, L., Lindstrom, F.T., Mc Farlane, C., and McCoy, E.L. (1988), Uptake of organic chemicals by plants: a theoretical model, *Soil Sci.,* 146: 403.
76. McCoy, E.L. (1987), Plant uptake and accumulation of soil applied trace organic compounds: theoretical development, *Agric. Syst.,* 25: 177.
77. Lindstrom, F.T., Boersma, L., and Mc Farlane, C. (1991), Mathematical model of plant uptake and translocation of organic chemicals: development of the model, *J. Environ. Qual.,* 29: 129.
78. Boersma, L., Mc Farlane, C., and Lindstrom, F.T. (1991), Mathematical model of plant uptake and translocations of organic chemicals: application to experiments, *J. Environ. Qual.,* 20: 137.

CHAPTER 4

Metabolic Processes for Organic Chemicals in Plants

Dieter Komoßa, Christian Langebartels, and Heinrich Sandermann, Jr.

TABLE OF CONTENTS

I. INTRODUCTION: GENERAL PRINCIPLES

Since the introduction of radiotracer and chromatographic techniques in the 1940s it has become increasingly clear that plants are able to metabolize organic chemicals that are foreign to them (xenobiotics). A useful early review[1] and many more recent reviews[2-9] summarize the data on intentionally applied xenobiotics (such as pesticides) as well as the numerous unintentionally released organic pollutants. Cell cultures have been of special value to demonstrate that plants can metabolize a great number of xenobiotics[10-14] ranging from highly polar xenobiotics such as glyphosate[15] to highly nonpolar chemicals such as DDT and hexachlorobenzene.[12]

Plant xenobiotic metabolism has been recognized to resemble liver metabolism of xenobiotics by a number of criteria.[10,16] One similarity is the above-mentioned broad range of metabolic substrates. In addition plant metabolism, like liver metabolism, can be divided conceptually into three phases, as summarized in Figure 1.

Phase I consists of transformation reactions which introduce functional groups(-OH, -NH$_2$, -SH) into various xenobiotic compounds (see Section II for recent case studies). Oxidation is the most frequently observed reaction type and often leads to detoxification or activation of a pesticide. Aromatic or alkyl hydroxylation, nitrogen or sulfur oxidation, epoxidation, as well as N- or O-dealkylation are major oxidative processes. Reductive reactions are less common and have been found for certain nitroaromatic compounds. Hydrolytic reactions may occur with carboxylic acid ester xenobiotics, organophosphorus compounds, carbamates, and anilides.

Phase II can be subdivided according to the "soluble" or "insoluble" ("bound") nature of the formed conjugates. The primary conjugation reactions are usually followed by processing reactions. This provides another similarity to liver metabolism of xenobiotics. The main reaction types for soluble conjugates include glucoside, glutathione, amino acid, and malonyl conjugation. The substituents are attached to functional groups existing in the xenobiotic or introduced in Phase I. Phenols and alcohols are commonly conjugated as β-O-D-glucosides. Xenobiotics that contain a primary or secondary amino group can be metabolized to N-glucosides. Glucose ester conjugates are formed with compounds containing carboxyl groups, while glutathione conjugation commonly involves displacement of halogen or nitro groups from the pesticide molecule. Xenobiotics can be further incorporated into biopolymers of the plant cell which are not soluble in the commonly used solvents ("insoluble" or "bound" residues). Many aromatic or heteroaromatic compounds with existing or introduced hydroxyl, carboxyl, amino, or sulfhydryl groups are known to be deposited into lignin or other cell wall components. A basic difference between plant and liver metabolic systems exists in phase III. Conjugates are predominantly excreted by aminals, whereas they are compartmentalized and stored within plant tissues.

Processing reactions through phases I to III are exemplified for two xenobiotics in Figure 2.[17] The herbicide 2,4-dichlorophenoxyacetic acid (2,4-D) is hydroxylated in phase I, conjugated to D-glucosyl and malonyl residues in phase II, and is compartmentalized in the plant vacuole as a soluble conjugate in phase III. The biocide pentachlorophenol (PCP) is hydroxylated and a chloro group is displaced in phase I. Tetrachlorocatechol is then conjugated to cell wall lignin (phase II) and is stored in the cell wall as an "insoluble" residue

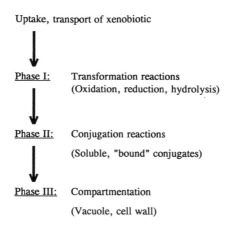

Uptake, transport of xenobiotic

Phase I: Transformation reactions
             (Oxidation, reduction, hydrolysis)

Phase II: Conjugation reactions

             (Soluble, "bound" conjugates)

Phase III: Compartmentation

             (Vacuole, cell wall)

FIGURE 1.   Plant metabolism of xenobiotics.

(phase III). The reactions shown are not yet completely documented. Furthermore, it is usually unclear whether the reactions of phases II and III occur simultaneously or sequentially.

The xenobiotic enzyme classes in plants at the same time resemble those of normal plant secondary metabolism and those of liver xenobiotic metabolism.[16] This is documented here by a list of purified plant isoenzyme species for xenobiotics and natural secondary substrates (Table 1).

For two of the enzyme classes shown in Table 1 (cytochrome P-450 monooxygenases and glutathione S-transferases) there is evidence for extensive DNA homologies which point to the existence of enzyme superfamilies reaching from bacteria to higher plants and man.[16] These similarities have strongly corroborated the previously proposed[10] designation of plant biomass as our "green liver" in addition to acting as our "green lung".

For modeling purposes, the plant metabolic processes of Figure 1 can be described by a series of partial reactions as shown in the minimum scheme of Figure 3. Each partial reaction is characterized by forward and back rate constants. Under steady-state conditions, the ratios of these rate constants represent equilibrium constants. However, each partial reaction is in reality likely to consist of numerous subreactions. True modeling of xenobiotic plant metabolism therefore requires a high kinetic resolution under carefully standardized experimental conditions. When a rate-controlling step (usually the slowest step) can be identified, the overall metabolism rate can be estimated from this step. In the case of transport-limited rates, estimations can be made from physicochemical data (see model section of this volume). The current status of metabolic modeling will be discussed in Section IV of this chapter.

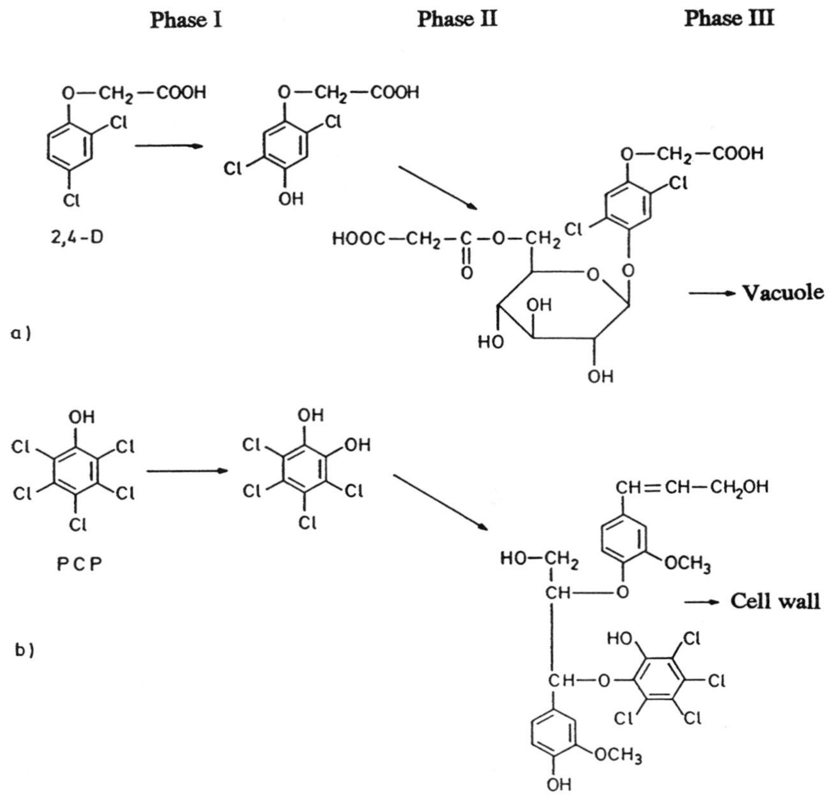

FIGURE 2. Plant xenobiotic metabolism of 2,4-dichlorophenoxyacetic acid (2,4-D) in soybean and pentachlorophenol (PCP) in wheat suspension cells. (Adapted from Reference 17.)

II. RECENT CASE STUDIES OF PLANT METABOLIC REACTIONS

A. EXAMPLES OF PHASE I REACTIONS

Among the phase I reactions oxidation, especially hydroxylation, plays an important role because the hydroxyl group is a suitable site for further conjugation. Increasing evidence points to the participation of cytochrome P-450 monooxygenases in the oxidation of several endogenous substrates[18] as well as several herbicides (Table 2). An amazing fact is that the P-450 substrate pesticides are in principle also substrates for the ubiquitous plant peroxidases,[30] but the peroxidase reactions are apparently not prominent *in vivo*. While this observation remains to be explained, reaction types such as hydroxylation, sulfoxidation, and *N*- and *O*-dealkylation have all been demonstrated to be

Table 1. Plant enzyme classes

Enzyme class	Xenobiotic substrates	Natural substrates
Phase I		
Cytochrome P-450 monooxygenases	4-Chloro-*N*-methylaniline	Cinnamic acid, pterocarpanes
Carboxylesterases	Diethylhexylphthalate	Lipids, acetylcholine
Phase II		
Glutathione *S*-transferases	Fluorodifen, alachlor, atrazine	Cinnamic acid
O-Glucosyltransferases	Chlorinated phenols	Flavonoids, coniferyl alcohol
O-Malonyltransferases	β-D-Glucosides of penta- chlorophenol and of 4-hydroxy-2,5-dichloro- phenoxyacetic acid	β-D-Glucosides of flavonoids and of isoflavonoids
N-Glucosyltransferases	Chlorinated anilines, metribuzin	Nicotinic acid
N-Malonyltransferases	Chlorinated anilines	1-Aminocyclopropylcarboxylic acid, anthranilic acid, D-amino acids

Note: Isoenzyme species that have been purified for xenobiotic substrates or for natural substrates, respectively, are listed. Table modified from Reference 16.

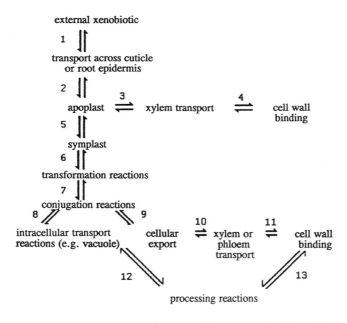

FIGURE 3. Simple kinetic model of plant metabolism as derived from Figure 1. The individual transport or metabolic processes are listed and numbered along with the arrows for the forward and back reactions. Each number shown is characterized by at least two kinetic constants as discussed in the text.

Table 2.   Cytochrome P-450-catalyzed hydroxylations of some herbicides

Herbicide	Plant	References
Chlorsulfuron	Wheat	19
Primisulfuron	Maize	20
Triasulfuron	Wheat	19
Diclofop	Wheat	19, 21, 22
Chlortoluron	Wheat Maize	23, 24 25
Bentazon	Sorghum Maize	26, 27, 28 29

Note: Arrows indicate the position of the hydroxylation.

cytochrome P-450 dependent. Cytochrome P-450 monooxygenase catalysis was recently documented for the N-demethylation of chlortoluron in wheat (*Triticum aestivum*),[23,24] the O-demethylation of metolachlor in sorghum (*Sorghum bicolor*),[28,31] and the N-demethylation of p-chloro-N-methylaniline in avocado (*Persea americana*).[32]

Chlortoluron, an N-dimethyl urea herbicide, can be metabolized in two different ways, namely by ring-methyl oxidation, typical for maize (*Zea mays*)[25] (Table 2), and/or by N-demethylation, e.g., in wheat.[23,24] A variety of other plants are also able to demethylate this herbicide. However, tolerance is only conferred on those species with the ability to didemethylate rapidly because the monodemethyl derivative is still phytotoxic. Cereal weeds, on the one hand, such as wild oat (*Avena fatua*), blackgrass (*Alopecurus myusuroides*), perennial ryegrass (*Lolium perenne*),[33] and Italian ryegrass (*L. multiflorum*)[34] slowly produce monodemethyl chlortoluron. Cotton (*Gossypium hirsutum*)[33-35] and lettuce (*Lactuca sativa*),[36] on the other hand, rapidly metabolize the herbicide further by sequential N-demethylation as well as ring-methyl hydroxylation.

Hydroxylation reactions are also documented by good recent examples of different tolerance due to different metabolic rates. Selectivity of sulfonylurea herbicides,[37,38] (e.g., chlorsulfuron, primisulfuron, and triasulfuron) has been correlated with the ability of cereal crops, maize, or soybean (*Glycine max*) to convert these substances rapidly to herbicidally inactive products. In contrast, susceptible weeds metabolize these herbicides much more slowly.[39,40] An interesting case for different metabolic pathways of crops and weeds is known for chlorsulfuron. Hydroxylation at the phenyl ring accounts for tolerance in wheat (Table 2), whereas hydroxylation at the methyl group of the triazine ring accounts for tolerance in flax (*Linum usitatissimum*) and black nightshade (*Solanum nigrum*).[9,41]

The ability of crop plants to protect themselves from herbicides by metabolism can be specifically enhanced by the use of safeners. Safeners such as naphthalic anhydride or benoxacor increase the selectivity of herbicides in the crop plant. They generally do not influence the detoxification in weeds.[42-44] A stimulation of specific enzymes such as cytochrome P-450 monooxygenases and glutathione *S*-transferases (see phase II reactions) could be demonstrated.

Two different mechanisms have been described for the detoxification of the herbicide bentazon, which is selectively used in several major crops, including soybean, cotton, and rice (*Oryza sativa*). In rice, maize, and sorghum, 6-hydroxybentazon and/or its glucoside have been found[26,29,45-48] (Table 2). Soybean, however, typically forms 6- and 8-hydroxybentazon.[49,50] In susceptible weeds such as velvetleaf (*Abutilon theophrasti*)[46,50] and shepherd's purse (*Capsella bursa-pastoris*)[51,52] no significant metabolites could be detected.

Hydroxylated derivatives are formed not only via direct hydroxylation by cytochrome P-450 monooxygenase-catalyzed reactions, but also by the cleavage of certain bonds. Oxidative cleavage of the ether bonds in the aryloxyphenoxypropanoate herbicides fenoxaprop ethyl,[53] quizalafop ethyl,[54] and CGA 184927[55] as well as gluthathione-mediated cleavage of the ether bonds in the diphenylether herbicides acifluorfen[56] and fluorodifen[57] provide examples of this reaction type.

Clomazone is a selective soybean herbicide. However, metabolic detoxification does not appear to account for tolerance since tolerant (soybean) and susceptible (velvetleaf) plants show the same metabolism pattern.[58] Oxidative cleavage of the clomazone molecule yielding 2-chlorobenzylalcohol followed by glucose conjugation is the main degradation reaction.[59]

Besides herbicide transformation, plants are also able to metabolize other xenobiotics (for reviews see References 4, 14, and 60 to 62). Even persistent chemicals such as the organochlorine insecticide DDT,[63] the fungicide hexachlorobenzene (HCB),[12] and the plasticizer chemical diethylhexylphthalate (DEHP)[64,65] are metabolized by plant cells. A turnover of polychlorinated biphenyls (PCBs) by cell cultures of *Rosa* cv. Paul's Scarlet has also been demonstrated.[66,67] Pentachlorophenol (PCP) has been used for a long time as a wood protectant. Tetrachlorocatechol, a potential mutagen, was identified as a primary metabolite of PCP in wheat cell cultures and plants[68] (Figure 2). In contrast, PCP was directly conjugated in soybean cells.[69]

Ester hydrolysis is important for the mode of action of the aryloxy-phenoxypropionate graminicides[70] because it represents bioactivation. The esters are used as herbicide precursors due to their better penetration of the cuticle. In the leaves, the esters are cleaved by carboxyl esterases. This is a common mechanism in tolerant and susceptible plants, but tolerant species rapidly metabolize the bioactive acids to nonphytotoxic products. For example, fenoxaprop ethyl was rapidly deesterified by wheat, barley (*Hordeum vulgare*), and crabgrass (*Digitaria ischaemum*), but only the weed was sensitive due to a very slow further metabolism of the free acid.[53] Similar results were found for diclofop methyl with resistant wheat and susceptible wild oat or annual ryegrass (*Lolium rigidum*).[71,72] However, different biotypes of *Lolium rigidum* vary extremely in their rates of metabolism of the free acid diclofop, and higher rates correlate with increased resistance.[72,73] On the other hand, three biotypes of *Avena fatua* differed in their tolerance to diclofop methyl, although the amounts of free dichlofop and the metabolic rates were very similar.[74]

A strong differentiation has to be made for the evaluation of the further metabolism of the deesterified aryloxyphenoxypropionate herbicides. Tolerance is generally marked by either the formation of hydroxy derivatives followed by glucosidation or, to a minor extent, by the cleavage of the ether bond. Diclofop is rapidly hydroxylated and conjugated with glucose in tolerant wheat and meadow grass (*Poa annua*).[75,76] The ether bonds in fenoxaprop and in the deesterified herbicide CGA 184927 are cleaved by wheat and barley,[53,55] and the ether bond in quizalofop, the free acid of quizalofop ethyl, is cleaved by soybean and cotton.[54] The formation of glycosyl ester conjugates of deesterified aryloxyphenoxypropanoates is generally considered not to be a true detoxifiction (see Section II.B).

For two sulfonylurea derivatives metabolic deesterification means inactivation of the herbicide. Thifensulfuron methyl is rapidly degraded to a single major metabolite, thifensulfuron acid, by soybean seedlings with a half-life of 4 to 6 h. Sensitive weeds such as redroot pigweed (*Amaranthus retroflexus*), common lamb's-quarters (*Chenopodium album*), and velvetleaf metabolize this herbicide much more slowly, with halflives greater than 36 h.[77] Soybean is also able to cleave the ester bond in chlorimuron ethyl, but the free acid chlorimuron, which is nonphytotoxic, represents only a minor metabolite.[78] The main metabolic route and detoxification lead to the homoglutathione conjugate (see Section II.B).

The selective use of herbicides is possible due to differences in phytotoxicity for crop and weed species. Broad-spectrum herbicides normally damage all plants. It was believed that nonselective herbicides, as a rule, could not be detoxified by plants.[79] Among the broad-spectrum herbicides, glyphosate (Roundup™) and phosphinothricin (Basta™) play important roles. Glyphosate interferes with the shikimate pathway in the biosynthesis of aromatic amino acids.[80] Phosphinothricin specifically inhibits a key emzyme in the nitrogen metabolism of the plant, glutamine synthase.[81] The selective application of this type of herbicide could be achieved by the introduction of plants with genetically engineered herbicide resistance.[82,83] Recent studies have documented the potency of plants to transform glyphosate[15] as well as phosphinothricin.[84] Cell suspension cultures of soybean, wheat, and maize were mainly used. Glyphosate was converted to the primary metabolite aminomethylphosphonic acid (AMPA), with soybean being most effective in this transformation reaction. AMPA was significantly incorporated into cell wall components, and in small amounts it was converted to unnatural phosphonolipids and CO_2. Four degradation products of phosphinothricin were detected in maize cell cultures (Figure 4).

B. EXAMPLES OF PHASE II REACTIONS

Phase II reactions include conjugation with D-glucose, glutathione, amino acids, or malonic acid to yield soluble conjugates.[85,86] Insoluble or "bound" conjugates are often formed with cell wall components such as lignin.[17] Typical examples for β-D-glucosides are provided by sulfonylureas or aryloxyphenoxypropionates (for references see above [phase I reactions], e.g., References 37, 38, and 70).

Glucosidation reactions are catalyzed by UDP-glucose-dependent glucosyltransferases. An O-glucosyltransferase for chlorinated phenols was isolated and highly purified from soybean cell suspension cultures.[87] Interestingly, two different O-glucosyltransferases isolated from soybean showed overlapping specificities for endogenous substrates (kaempferol and p-hydroxyphenylpyruvic acid, respectively) and the herbicidal metabolite 6-hydroxybentazon.[88] Due to the fact that the Michaelis constants for both natural compounds are about 20-fold smaller than those for 6-hydroxybentazon, it was concluded that the primary role of these enzymes involves the conjugation of endogenous metabolites.

Xenobiotics with free carboxyl groups can be coupled with glucose to glucose esters. Recently, a specific UDP-glucose-dependent glucosyltransferase for 2,2-bis-(4-chlorophenyl)acetic acid (DDA), the main catabolite of DDT,[63] was isolated and purified from soybean cell suspension cultures (A.M.L. Wetzel and H. Sandermann, 1992, unpublished results). A screening program revealed enzymatic activity toward DDA in a number of plant species. The formation of glucose ester conjugates of deesterified aryloxyphenoxypropanoate herbicides appears only in susceptible plants, where the herbicidally active acid can be easily liberated again.[8,9,70] This is true for weeds such as cockspur grass (*Echinochloa crus-galli*), Italian ryegrass, annual ryegrass, and wild oat which form glucose esters of diclofop.[8,9,72,73,76] Maize which is highly sensitive to CGA 184927 converts the corresponding carboxylic acid to glycosyl ester conjugates.[55] It appears that different metabolic rates of herbicides do not always confer tolerance to a plant.

The formation of N-glucosides also occurs in plants, although it is less common. N-Glucosidation represents a major detoxification step in the metabolism of the *asym*-triazinone herbicide metribuzin in tomato (*Lycopersicon esculentum*), whereas it is only of minor importance in soybean.[89] The

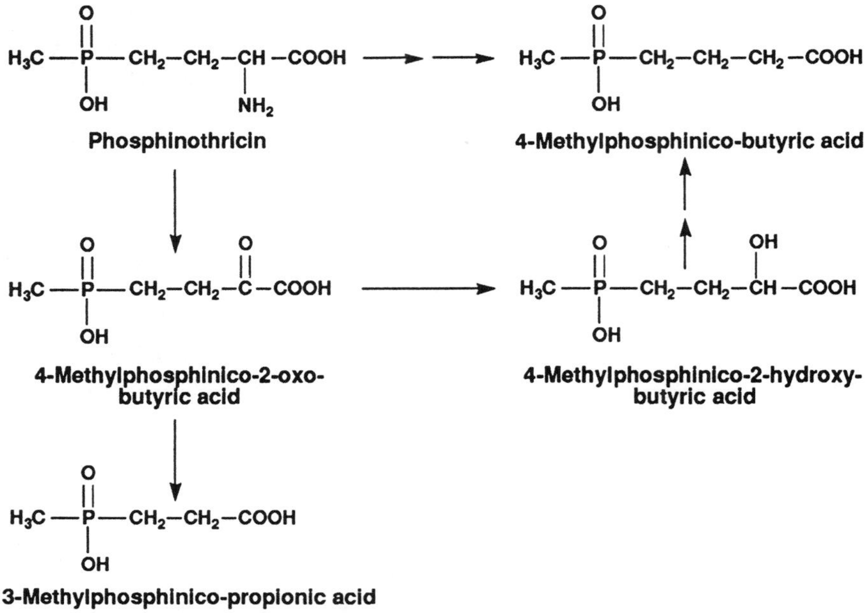

FIGURE 4. Proposed scheme of phosphinothricin metabolism in maize cell suspensions cultures. (From Komoßa, D. and Sandermann, H. (1992), *Pestic. Biochem. Physiol.*, 43: 95. With permission.)

N-glucosyltransferase for metribuzin was isolated from tomato cell cultures and partially characterized.[90] Cell cultures of soybean were a suitable source for the isolation of a *N*-glucosyltransferase for chlorinated anilines.[87,91]

Glycosides of natural and xenobiotic substrates are often processed by coupling with other carbohydrate moieties or by acylation. Examples for oligosaccharide conjugates are given by the xylosylglucoside of *p*-nitrophenol in cell cultures of periwinkle (*Catharanthus roseus*)[92] and by the rhamnosylglucosides of phenylacetonitrile oxim and benzyl alcohol, two degradation products of the insecticide phoxime, in cell cultures of *Chenopodium rubrum*.[93] Acylation of glucosides, typically at the 6-OH group with malonic acid via malonyl-CoA, leads to pesticidal end products which are believed to be stored in the vacuole or to be excreted into the apoplast.[17,94] The aforementioned glucosides of metribuzin[89] as well as of PCP[69] are malonylated. Besides the formation of the disaccharide conjugate of *p*-nitrophenol, the 6-*O*-malonyl-*O*-β–D-glucoside has also been described in *Catharanthus roseus* cell cultures.[94] Malonylated glucosides are also common plant metabolites of endogenous substrates.[95] Such acyl conjugates are alkali labile and may have been overlooked in a variety of investigations on glycoside conjugation of pesticides.

Besides the formation of O-malonyl derivatives, N-malonyl conjugates are also formed from endogenous substrates or xenobiotics. D-Amino acids and 1-aminocyclopropane-1-carboxylic acid (ACC), the biosynthetic precursor of the plant hormone ethylene, are natural substrates.[96] Four distinct N-malonyltransferases with different substrate specificities were isolated from peanut (*Arachis hypogaea*), one being active with 3,4-dichloroaniline (Table 1). The main degradation product of the insecticide phoxime in soybean plants and cell cultures was the N-malonyl conjugate of phenylacetonitrile amine.[97] The corresponding N-malonyltransferase, purified to apparent homogeneity, showed activity with 4-chloroaniline and to a lesser extent with neutral D-amino acids.[94] A further N-malonyl conjugate was described in the metabolism of metribuzin in soybean, but the N-malonyl derivative of dethiomethyl metribuzin was only a minor degradation product.[89] Malonylation has been considered to be a final step in xenobiotic metabolism.[85,94] Again the question arises whether the metabolic conversions are catalyzed by highly specific enzymes or by enzymes with broad substrate specificities for natural and xenobiotic compounds. Until now there have been few investigations on plant enzymes involved in xenobiotic metabolism that have yielded a preliminary answer.

Conjugation with glutathione (GSH) plays a central role as a detoxification mechanism for herbicides in plants.[85,86,98] The most famous example is given by the resistance of maize to atrazine.[71] Some leguminous species (e.g., soybean) contain very little GSH and instead form conjugates with homoglutathione (hGSH). Conjugation with (homo)glutathione is catalyzed by glutathione *S-transferases* (GSTs) and can occur by distinct reaction types (Table 3).

The detoxification via glutathione belongs to the best investigated and most intensively characterized areas of herbicide metabolism. Safeners have been shown to either increase the level of glutathione or increase the activity of certain glutathione S-transferase activities.[42] Some GST enzymes from maize have been studied in detail. GST I, II, and III have similar molecular weights (about 50 kDa as dimers). The three isoenzymes, with GST II only inducible by safener treatment, showed different substrate specificities for alachlor, metolachlor, and 1-chloro-2,4-dinitrobenzene, but little activity for atrazine. It is assumed that there are two additional GSTs for conjugation of atrazine.[107] Glutathione conjugates undergo further complex metabolism by numerous reactions. As an example, processing the primary glutathione conjugate of fluorodifen in spruce (*Picea abies*) cell cultures[57] is shown in Figure 5.

"Bound" conjugate fractions have remained difficult to characterize. A suitable method for cell wall fractionation was published by Langebartels and Harms in 1985.[108] Plant cell wall material was degraded in a stepwise manner (enzymatically and chemically) to yield six fractions which were operationally defined to be mainly composed of starch, protein, pectin, lignin, hemicellulose, or cellulose.[108] PCP-bound residues (>30% of the applied substrate) were mainly associated with the hemicellulose fraction in wheat cell cultures.[108]

Table 3.   **Different types of herbicidal conjugation reactions catalyzed by glutathione *S*-transferases**

Reaction type of (h)GSH conjugation	Herbicide class	Examples (herbicide/plant)	References
Nucleophilic substitution of a chlorogroup	*sym*-Triazine	Atrazine/maize	71
	Chloroacetamide	Propachlor/soybean[a]	99
		Pretilachlor/rice	100
		Metazachlor/maize	101
		Acetochlor/maize soybean[a]	102
		Metolachlor/maize	103
	Sulfonylurea	Chlorimuron/soybean[a]	78
		maize	104
Cleavage of a diphenyl ether bridge	Diphenylether	Fluorodifen/spruce	57
		Acifluorfen/soybean[a]	56
Fission of a sulfoxide	Thiocarbamate	EPTC/maize cotton	104
	asym-Triazinone	Metribuzin/soybean[a]	89
Addition to an epoxide		Tridiphane/maize	105
		giant foxtail	106

Note: The (h)GSH conjugates with chloroacetamides also form nonenzymatically.

[a] Soybean always forms the hGSH conjugate.

Lignin binding of PCP has been characterized by a number of methods, including ^{13}C-nuclear magnetic resonance (NMR) spectroscopy.[68] Feeding of ^{14}C-labeled 2,4-dichlorophenol or 4-chloroaniline to maize cell cultures led to the formation of 50 and 40% bound residues, respectively, with the protein, pectin, and lignin fractions containing major portions of ^{14}C activity.[109] Treatment of intact wheat plants with [U-^{14}C]-3,4-dichloroaniline (DCA) led to about 60% deposition into the insoluble residues. The sequential solubilization procedure revealed that about 85% of the ^{14}C label was associated with the operationally defined lignin fraction.[110] Two types of linkage between DCA and lignin were proposed on the basis of animal bioavailability and acid hydrolysis experiments[110] (Figure 6). The labile linkage shown was proposed to explain an unusual "burst" release upon treatment of the lignin metabolite fraction under stomach conditions (0.1 *N* HCl, 37°C).[110] A binding of pesticide metabolites to lignin-like material is prominent in many of the published reports concerning insoluble residues.[111-113]

C. EXAMPLES OF PHASE III REACTIONS

Phase III of the plant xenobiotic metabolism involves storage and compartmentation of soluble conjugates in the vacuole and of insoluble conjugates in the cell wall. The unequivocal localization of a soluble pesticidal conjugate in the vacuole could be demonstrated only in the case of 2,4-D[14] (Figure 2). Several plant secondary products are known to be major constituents of the vacuole.[115] It is assumed that xenobiotic 6-*O*-malonyl-β-D-glucosides are often stored in the vacuole. Malonylation possibly serves as a signal

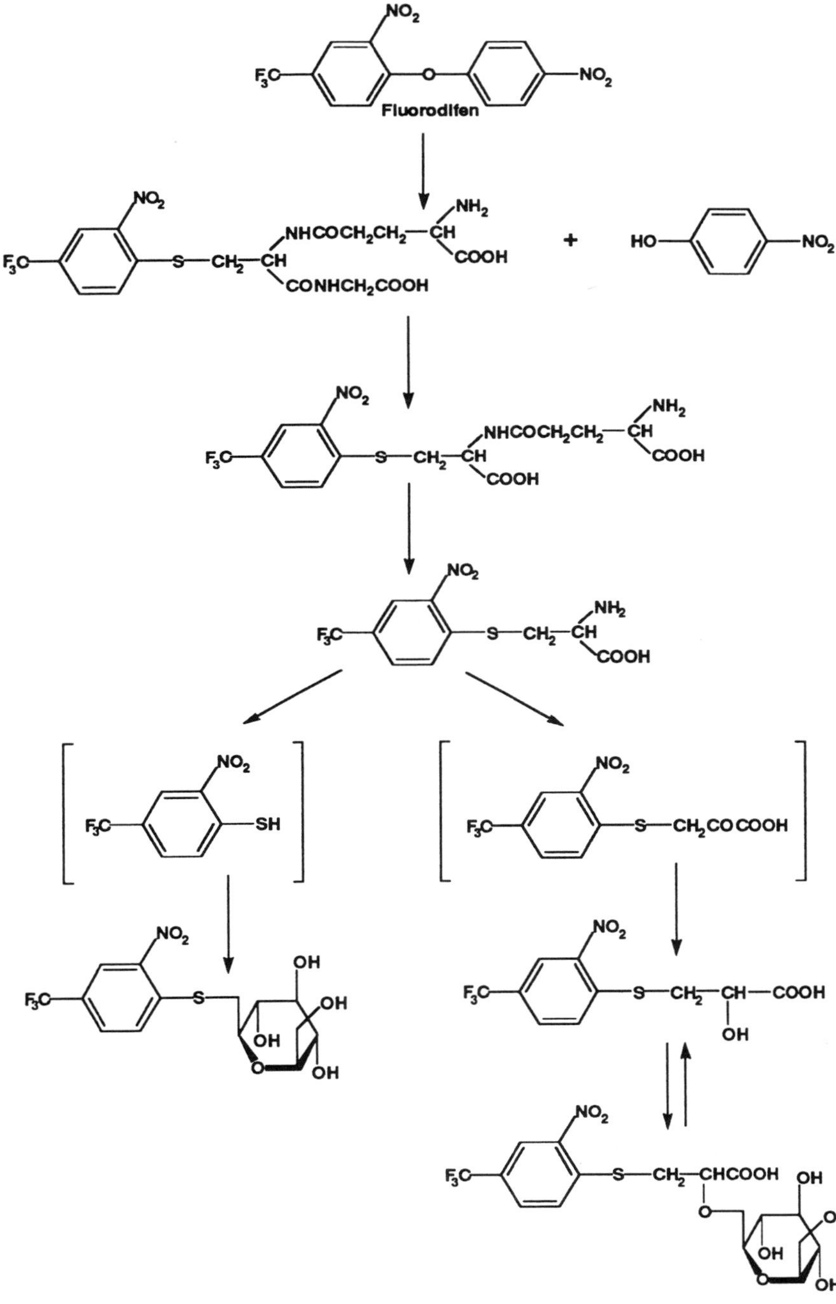

FIGURE 5. Proposed metabolism of fluorodifen in spruce cell suspension cultures via glutathione conjugation. (Adapted from Reference 57.)

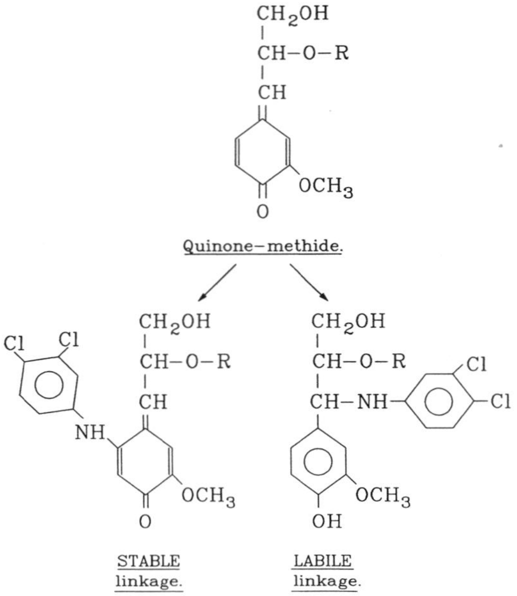

FIGURE 6. Lignin linkages of 3,4-dichloroaniline. (From Sandermann, H., Musick, T.J., and Aschbacher, P.W. (1992), *J. Agric. Food Chem.*, 40:2001. With permission.)

for deposition of the substance in the vacuole[69] or for export into the apoplast.[16] Some malonylglucosyl conjugates of herbicides have already been identified (see Section II.B). However, it is likely that further conjugates may have been overlooked due to the instability of the malonyl linkage and instead were incorrectly identified as simple glucosides.[86,86] The localization of xenobiotic metabolites in the vacuole is difficult due to technical problems in isolating intact vacuoles. Therefore, it is not possible at present to attribute compartmentation of pesticidal conjugates in the cellular vacuole to resistance mechanisms.[116] In the case of "bound" residues no clear assignment to cellular compartments has been published so far. It is commonly assumed that the cell wall is the major site of deposition.

III. STANDARDIZED SYSTEMS FOR XENOBIOTIC METABOLISM IN PLANTS

A. STUDY SYSTEMS

Standardized model systems to study the metabolic fate of environmental chemicals in terrestrial plants have been used for more than 20 years.[117,118] Their scope (i.e., complexity) ranges from laboratory systems with defined biotic and abiotic components representing general processes to microcosm and field experiments with a temporal scale of weeks to months. An outdoor lysimeter system has been successfully used by Führ and coworkers to determine the long-term fate of ^{14}C-labeled herbicides in plants and soil.[119,120] The system is performed under "agricultural management practices" and includes a realistic simulation of climate, soil, and biotic parameters. The long-term fate of compounds during one or two vegetation periods can be determined in the system but, due to the high expenditure for outdoor application of radiolabeled chemicals, the system is not designed to compare a large number of chemicals. A laboratory model ecosystem was developed by Schuphan et al.[121] which allows a quantitative evaluation of the metabolic fate of test compounds, including volatile metabolites. Phytotoxicity and metabolism studies were performed in standardized systems with various plant species by Mc Farlane et al.[122] The uptake of organic compounds from the soil or from the air was studied in the group of I. Scheunert.[123,124] The data on the distribution of ^{14}C label are used to predict the chemical concentrations in plants growing on contaminated soil.[125]

Undifferentiated plant cell cultures in liquid media represent a means by which metabolic transformations of xenobiotics may be attributed unequivocally to plant, rather than microbial, metabolism. Rates of transformation are often more rapid than in intact plants. Cell suspension cultures can be obtained for all major agricultural crops. This allows one to study the fate of a given pesticide in any target plant species. Comprehensive reviews of xenobiotic metabolism in plant cell cultures have appeared.[10-12,14,94,126,127] Other uses of plant cell cultures in phytotoxicity studies for xenobiotics and on the mode of action of plant growth regulators have been summarized by Gressel[128] and Grossmann.[129]

B. THE STANDARDIZED CELL CULTURE METABOLISM TEST

In the early 1980s, a standardized laboratory test with suspension cultures of soybean and wheat was developed by three groups in Germany.[12,13,130] The cell culture metabolism test was used to study the metabolism of ^{14}C-labeled xenobiotics both in the cells and in the culture medium following an incubation period of 48 h. To date over 80 chemicals have been assayed with this test procedure. It is now routinely used in several laboratories and will be incorporated into the German guidelines for testing of chemicals at the second stage level.[131] In detail, the system

- Provides a rapid overview of the main cellular processes, formation of less polar compounds, polar conjugates (mainly phase II soluble conjugates), and nonextractable residues (phase II "bound" conjugates)
- Yields data on the metabolic capacity of the plant cells proper because microbial contamination and photochemical degradation can be excluded
- Points to major differences between dicot and monocot species
- Can be used as a standard screening test for newly developed compounds

The plant cultures used are highly standardized heterotrophic cell suspensions which have now been propagated for more than 20 years. The soybean culture (*Glycine max* L. Merr. cv. Mandarin) was initiated from root meristems in 1964.[132] The wheat culture (*Triticum aestivum* L. cv. Heines Koga II) was established from root sections.[133] Both cultures are grown on liquid B5 medium containing 2,4-D as the sole growth hormone. They are cultured in Erlenmeyer flasks at 27°C on a gyratory shaker in the dark. Since their initiation, the cultures have passed more than 1200 (soybean) or 500 (wheat) passages and have retained their typical growth behavior. The cultures have been included in the German Collection for Microorganisms and Plant Cell Cultures (Dr. Schumacher, DSM, Mascheroder Weg 1B, D-38124 Braunschweig, Germany).

The procedure for the cell culture metabolism test is summarized in Figure 7. Details of the test performance can be found in Reference 130. The test compounds are supplied to five to ten parallel cultures at the end of their logarithmic growth stage at a concentration of 1 mg/l (1 ppm). The cultures are incubated for 48 h in the dark and then checked for microbial contamination (usually >98% aseptic). The cells are separated from the medium by filtration under mild vacuum and are washed with distilled water. Cells and liquid culture media are extracted in methanol/chloroform (Bligh-Dyer extraction) or methanol/dichloromethane, both at 2:1 (v/v). The cells are then homogenized by ultrasonic disintegration. Organic and aqueous phases are analyzed by thin layer chromatography on precoated silica gel plates. The solvent systems are designed in a way that the parent compound has an Rf of about 0.5. Metabolite fractions with lower Rf are termed "less polar" metabolites, while those with higher Rf are termed "polar" compounds. The major metabolites are then further purified by high-pressure liquid chromatography (HPLC) and characterized by mass spectrometry (MS) and NMR spectroscopy. The "insoluble" residues after Bligh-Dyer extraction are combusted and their radioactivity is determined by liquid scintillation counting. The residues can be further analyzed by cell wall fractionation.[108]

The detailed results for 28 compounds with different physicochemical properties are given in Table 4 in the order of increasing *n*-octanol/water partitioning coefficients. The reader is referred to the references given in the table for metabolite identification . Examples for metabolic pathways analyzed in the standard cell culture metabolism test are given in Figures 2, 4, and 6. The main results for the compounds of Table 4 are as follows:

- All of the test chemicals were metabolized, with mean metabolic rates of 45 and 39% in soybean and wheat cultures, respectively. This was also evident for compounds which were regarded as nondegradable in intact plants, e.g., glyphosate, phosphinothricin, benzo(a)pyrene, and DDT. It can be deduced that the corresponding enzyme systems exist in plant cells. Apparently, all compounds can be transformed in plants provided that transport and penetration barriers do not become rate

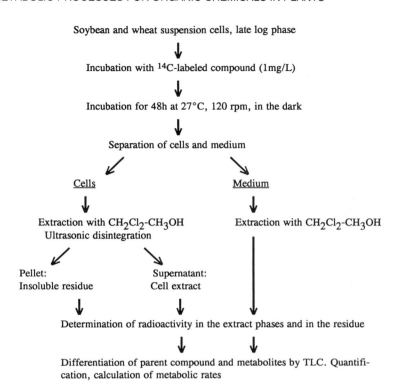

Soybean and wheat suspension cells, late log phase

Incubation with ^{14}C-labeled compound (1mg/L)

Incubation for 48h at 27°C, 120 rpm, in the dark

Separation of cells and medium

Cells   Medium

Extraction with CH_2Cl_2-CH_3OH Extraction with CH_2Cl_2-CH_3OH
Ultrasonic disintegration

Pellet: Supernatant:
Insoluble residue   Cell extract

Determination of radioactivity in the extract phases and in the residue

Differentiation of parent compound and metabolites by TLC. Quantification, calculation of metabolic rates

FIGURE 7. Schematic outline of the incubation and work-up procedure of the cell culture metabolism test. TLC = thin layer chromatography.

limiting. General physicochemical properties of the test compounds such as the *n*-octanol/water partitioning coefficient, water solubility, and molecular weight were not correlated with their metabolic fate in suspension cells. Instead, transformation types correlated with existing or introduced functional groups of the molecules, as detailed in Section I. The metabolism data for a given compound were similar to within ±10% when the test was performed by different laboratories. This defines the high degree of reproducibility of the cell culture metabolism test.

• Polar conjugates were the main metabolite fraction for most of the xenobiotics tested. The degree of conjugate formation was comparable in both cultures (ranging from 1.1 to 91% in soybean [means of 37%] and from 0.3 to 89% in wheat [means of 26%] of the applied radioactivity). The conjugation types, however, were different in the two cultures, with *N*-malonyl and *O*-malonyl conjugates prevailing in soybean and *O*- and *N*-glucosides prevailing in wheat cells (see Section II).

Table 4. Metabolism of xenobiotics in soybean (S) and wheat (W) suspension cultures under the experimental conditions of the standardized plant cell culture metabolism test

Compound	Log K_{ow}[a]	Culture	Parent comp.	Metabolites		Residue	Metabolic rate	Total	Ref.
				Polar	Nonpolar				
Diquat	−4.6	S	68.0	2.7	1.7	21.3	25.7	93.7	134[b]
		W	63.6	0.3	2.9	21.6	24.8	88.4	134
Glyphosate	0.00	S	40.6	57.6	0.0	1.7	59.3	100.0	15
		W	97.0	2.7	0.0	0.7	3.4	100.4	15
Phosphinothricin	0.10	S	96.7	2.4	0.0	0.1	2.5	99.2	84
		W	98.4	1.1	0.0	0.1	1.2	99.6	84
Metaldehyde	0.12	S	80.2	5.3	4.8	0.6	10.7	90.9	134
		W	79.0	5.8	5.2	1.4	12.4	91.4	134
Dimethoate	0.74	S	62.0	30.7	1.0	2.9	34.6	96.6	134
		W	84.9	7.7	3.5	0.5	11.7	96.6	134
Aldicarb	1.14	S	2.4	91.2	0.7	1.7	93.6	96.0	134
		W	8.3	88.9	0.4	0.5	89.8	98.1	134

Compound										
4-Chloroaniline	1.83	S	16.6	72.2	0.0	3.8	76.0	92.6	13
		S	6.7	91.8	0.0	2.3	94.1	100.8	135
		W	6.9	14.9	0.0	72.2	87.1	94.0	13
		W	13.5	26.0	0.0	66.9	92.9	106.4	135
Monolinuron[c]	2.20	S	75.6	19.0	0.7	3.0	22.7	98.3	134
		W	88.8	8.3	0.1	2.4	10.8	99.6	134
Atrazine	2.34	S	14.8	72.5	0.4	7.6	80.5	95.3	134
		W	76.6	12.4	0.4	1.1	13.9	90.5	134
3,4-Dichloroaniline	2.69	S	8.8	79.1	0.0	2.5	81.6	90.4	13
		S	2.4	102.1	0.0	2.7	104.8	107.0	135
		W	10.9	51.6	0.0	31.0	82.6	93.5	13
		W	10.4	62.0	0.0	31.3	93.3	103.7	135
2,4,5,-Trichlorophenoxyacetic acid	2.79	S	92.6	11.3	0.0	2.2	13.5	106.1	12
		S	81.9	7.7	2.4	1.2	11.3	93.2	134
		W	77.8	7.0	0.6	1.5	9.1	86.9	12
		W	46.9	23.0	19.7	3.2	45.9	92.8	134
2,4-Dichlorophenoxyacetic acid	2.81	S	76.0	11.9	0.2	2.0	14.1	90.1	13
		S	80.1	17.2	0.8	3.5	21.5	101.6	134
		W	7.5	69.7	4.0	9.4	83.1	90.6	13
		W	9.0	68.5	6.6	17.7	92.8	101.8	134

Table 4. Metabolism of xenobiotics in soybean (S) and wheat (W) suspension cultures under the experimental conditions of the standardized plant cell culture metabolism test (continued)

Compound	Log K_{ow}[a]	Culture	Parent comp.	Metabolites Polar	Metabolites Nonpolar	Residue	Metabolic rate	Total	Ref.
Carbaryl	2.81	S	33.2	39.1	0.9	22.4	62.4	95.6	134
		W	57.4	24.8	2.2	11.3	38.3	95.7	134
Malathion	2.89	S	4.2	76.5	0.4	11.4	88.3	92.5	134
		W	2.0	82.5	5.2	6.8	94.5	96.5	134
Chlorpropham	2.92	S	15.9	51.2	0.8	26.0	78.0	93.9	134
		W	48.0	24.1	1.2	22.9	48.2	96.2	134
Triadimenol	3.08	S	76.4	16.7	2.6	2.2	21.5	97.9	134
		W	72.7	15.8	1.4	2.1	19.3	92.0	134
Methoxychlor	3.31	S	40.7	31.1	4.0	17.6	52.7	93.4	134
		W	56.4	33.6	0.8	2.9	37.3	93.7	134
Lindane	3.76	S	77.7	9.6	0.5	1.1	11.2	88.9	134
		W	78.8	7.4	1.9	0.6	9.9	88.7	134

Compound		S/W							
Parathion	3.83	S	15.2	73.0	0.8	3.6	77.4	92.6	134
		W	4.7	54.1	0.2	38.7	93.0	97.7	134
Diflubenzuron	3.89	S	76.8	6.5	1.8	6.4	14.7	91.5	134
		W	81.4	7.3	1.6	1.7	10.6	92.0	134
Endosulfan	4.7	S	5.4	83.9	0.1	1.6	85.6	91.0	134
		W	45.8	48.2	0.2	2.3	50.7	96.5	134
Pentachlorophenol	5.01	S	21.1	45.3	0.2	6.3	51.8	72.9	12
		S	25.4	61.8	0.0	11.3	73.1	98.5	12
		S	33.3	53.1	0.2	6.9	60.2	93.5	134
		S	27.6	59.9	0.2	7.1	67.2	94.8	13
		W	3.0	51.6	0.1	37.6	89.3	92.3	12
		W	1.0	71.6	0.3	18.7	90.6	91.6	134
		W	1.3	54.0	0.0	41.3	95.3	96.6	13
Diethylhexylphthalate	5.03	S	74.7	26.3	0.0	0.6	26.9	101.6	13
		W	60.1	23.3	1.7	3.7	28.7	88.8	65
		W	55.4	24.3	0.4	7.7	32.4	87.8	13
Hexachlorobenzene	5.47	S	84.5	2.3	0.0	2.2	4.5	89.0	12
		W	33.1	29.4	0.0	3.7	33.1	66.2	12

Table 4. Metabolism of xenobiotics in soybean (S) and wheat (W) suspension cultures under the experimental conditions of the standardized plant cell culture metabolism test (continued)

Compound	Log K_{ow} [a]	Culture	Parent comp.	Metabolites		Residue	Metabolic rate	Total	Ref.
				Polar	Nonpolar				
DDE	5.69	S	102.4	1.1	0.0	0.9	2.0	104.4	63
		W	95.3	1.8	0.0	3.7	5.5	100.0	63
Benzo(a)pyrene	6.04	S	19.4	49.7	7.5	15.6	72.8	92.2	13
		W	51.7	22.7	5.8	9.0	37.5	89.2	13
DDT	6.19	S	84.4	1.4	5.3	2.4	9.1	93.5	63
		W	95.8	2.4	0.0	2.3	4.7	100.5	63
Terylene	6.80	S	27.9	52.5	0.3	8.5	61.3	89.2	13
		W	80.6	6.9	0.0	3.0	9.9	90.5	13

[a] Data from References 13 and 136.

[b] Test compounds from Reference 134 with apparent transformation rates of more than 12% in heat-inactivated control cultures are not included in this table.

[c] Modified test procedure.

- Mean amounts of 6% (soybean) and 11% (wheat) of the applied radiolabel were found in the so-called insoluble residues (phase II "bound" conjugates) which consist of various macromolecules, mainly cell wall lignin and polysaccharides. Bound residue formation was especially high for chemicals with aromatic and heteroaromatic ring structures. Wheat cells showed a significantly higher incorporation into bound residues as compared to soybean, and radiolabel in this fraction reached 31% (3,4-dichloroaniline), 39% (parathion), and 70% (4-chloroaniline). Cell wall binding of xenobiotics may be a typical process for monocot cell cultures, since maize and barley cultures also showed higher residue formation than dicot cells.[113] In addition, wall-bound residues of 4-chloroaniline and 2,4-dichlorophenol were higher in maize than in tomato suspension cells under experimental conditions similar to those above.[109]
- Less polar metabolites were found as a minor type of transformation (range of 0 to 7.5%). Higher amounts were only detected for 2,4,5-trichlorophenoxy-acetic acid (2,4,5-T) in wheat (Table 4). The major metabolite was identified as 2,4,5-trichlorophenol.[134] The high recovery rates for the applied radiolabel (mean values of 95 and 94% for soybean and wheat, respectively) suggest that mineralization to volatile CO_2 was not a major factor for any of the test compounds. This was confirmed by a modified test procedure[137] in which volatile compounds were trapped by polyurethane foam plugs and CO_2 absorption mixtures.
- The polar metabolites were mainly cell associated, but in soybean suspensions a considerable part of the radiolabel was found in the culture medium, e.g., 39% for benzo(a)pyrene, 30% for pentachlorophenol, 59% for 4-chloroaniline, and 64% for 3,4-dichloroaniline.[13,69,135] The percentage of medium-localized metabolites of seven xenobiotics was markedly higher (35 ± 9%) in soybean than in wheat cultures (5 ± 1%).[13] This was partly due to the rapid formation and secretion of malonyl conjugates,[69,135,138] which are known to be directed toward either vacuoles or the extracellular space.[135]

It was concluded from the study on 23 pesticides by Schmidt et al.[134] that the test system allowed the classification of the persistence of the compounds. Soybean and wheat cultures showed similar turnover rates for 16 compounds, while transformation differed by more than 30% for the remaining 7 chemicals (Table 4). A comparison of *p*-nitrophenol metabolism in heterotrophic, photomixotrophic, and photoautotrophic *Catharanthus roseus* cells showed no qualitative or quantitative differences in metabolite formation.[94] "Green" and "nongreen" cultures of *Chenopodium rubrum*, however, differed in phase I reactions for the insecticide phoxime, while conjugate formation was comparable. Results of the cell culture metabolism test have been compared with those from intact plants.[113,138,139] The metabolism pattern was comparable for more than 20 xenobiotics, but marked quantitative differences were sometimes observed, especially with nonpolar xenobiotics. Soluble conjugates and "bound" residues also represented the major metabolite classes in intact plants. The rates of bound residue formation were usually higher in intact plants.

It can be concluded that the cell culture metabolism test is able to provide a reasonable approximation of the metabolism pattern expected with intact young plants. The test can be performed with ease and rapidity within 1 to 2 weeks with ten parallel cultures. Furthermore, the metabolic rates of the cultures are high enough to allow an identification of the major metabolites for new synthetic compounds.

IV. CONCLUSIONS

The above text documents with many examples that plants have a high metabolic capacity for a variety of polar as well as nonpolar xenobiotics. For modeling purposes, it may be useful to begin with two extreme situations: (1) rate limitation of metabolism by the cuticle or other surface structures and (2) rate limitation of metabolism by internal metabolic enzymes. The standardized plant cell cultures described in Section III are devoid of cuticles, and metabolic rates do not correlate with the octanol/water partitioning coefficients. This conclusion had been drawn previously[12,134] and is documented here by the data of Table 4. In contrast, metabolic systems where the plant cuticle or root epidermis is rate limiting show a strong negative influence of the octanol/water partitioning coefficient. This is described in detail in Chapters 3 and 6 of this volume.

In the cases of pentachlorophenol and 3,4-dichloroaniline in soybean cell suspension cultures and in the case of diethylhexylphthalate in wheat cell suspension cultures, the initial metabolic plant enzymes have been highly purified and characterized. Their relevant kinetic data are summarized in Table 5. It had been shown previously that the metabolism of DEHP in wheat cell suspension cultures differs greatly from that in the intact plant. Metabolism was apparently limited either by partitioning into cuticles or lipid droplets or by absorption to cell walls in the intact plants.[65] In contrast, in the cell culture system the initial enzyme, DEHP esterase, appeared to be rate limiting. This is documented here by the finding that the ratio of observed cellular reaction rate to *in vitro* rate was close to one (Table 5). Similar calculations have been made for the initial enzymes for soybean PCP metabolism (through *O*-glucosyltransferase activity) and 3,4-dichloroaniline metabolism (through *N*-malonyltransferase activity). As shown in Table 5, in these cases the *in vivo* metabolic rates appeared to be lower by a factor of 5 to 10 than enzyme capacity estimated for the *in vivo* test. The differences may, however, not be relevant in view of the crude assumptions made. Table 5, therefore, can be considered to contain the first indications for a rate limitation of cellular

Table 5. Comparison between cellular metabolism rates and expected activity of the first enzyme of metabolism for xenobiotics

Substance/culture first enzyme	DEHP/wheat DEHP esterase	PCP/soybean O-glucosyltransferase	DCA/soybean N-malonyltransferase
Cellular reaction rate to polar metabolites (% of applied; nmol/40 ml)[a]	29; 29	55; 82.5	79.1; 194.6
Same, expressed as nkat/kg protein[a]	10.4	14.0	33
Enzyme rate in crude *in vitro* extract (nkat/kg] V_{max})	45.5 [140]	960 [87]	32,000 [87]
K_m value to reach 50% of V_{max} (μM)	5 [140]	30 [87]	100 [87]
Concentration (μM) in cellular experiment; % of K_m	2.5; 50	4.8; 16	2.1; 2.1
% of V_{max} expected in cellular experiment;[b] nkat/kg protein	25; 11.3	8; 77	1; 320
Ratio of observed cellular reaction rate over *in vitro* rate	0.92	0.18	0.10

Note: Numbers in brackets are references. DEHP = diethylhexylphthalate; PCP = pentachlorophenol; DCA = dichloroaniline.

[a] 1 nkat/kg protein = 1 nmol/kg protein. The 4 g of wheat cells used contain 16.3 mg soluble protein,[140] while the 8 g of soybean cells used contain 34 mg soluble protein.[87]

[b] It is assumed that there is linearity between reaction rate and initial substrate concentration in the range below K_m. Substrate depletion is neglected.

metabolism by intracellular enzyme activities for the case of suspension-cultured plant cells.

A preliminary mathematical model to describe the metabolism of atrazine in barley plants also has been elaborated.[141] However, more generally it can be stated that no detailed mathematical treatment of xenobiotic plant metabolism is so far available. The resolution of plant metabolic studies has been too low to discern individual metabolic rate or equilibrium constants according to Figure 3. In contrast, a high degree of resolution has been achieved in modeling bioconcentration and transport of organic substances (see Chapters 3 and 6).

ACKNOWLEDGMENTS

The authors would like to thank Lucia Gößl, Hanna Wegner, and Robert G. May for help in preparing the manuscript and figures. Work from this laboratory has been supported by the Bundesministerium für Forschung und Technologie, Deutsche Forschungsgemeinschaft, and the Fonds der Chemischen Industrie.

REFERENCES

1. Casida, J.E. and Lykken, L. (1969), Metabolism of organic pesticide chemicals in higher plants, *Annu. Rev. Plant Physiol.,* 20: 607.
2. Naylor, A.W. (1976), Herbicide metabolism in plants, in *Herbicides, Physiology, Biochemistry, Ecology,* Vol. 1, L.J. Audus, Ed., Academic Press, New York, p. 397.
3. Shimabukuro, R.H. and Walsh, W.C. (1979), Xenobiotic metabolism in plants: In vitro tissue, organ, and isolated cell techniques, in *Xenobiotic Metabolism: In Vitro Methods,* ACS Symposium Series 97, G.D. Paulson, D.S. Frear, and E.P. Marks, Eds., American Chemical Society, Washington, D.C., p. 3.
4. Shimabukuro, R.H., Lamoreux, G.L., and Frear, D.S. (1982), Pesticide metabolism in plants, reactions and mechanisms, in *Biodegradation of Pesticides,* F. Matsumura and C.R.K. Murti, Eds., Plenum Press, New York, p. 21.
5. Hatzios, K.K. and Penner, D. (1982), *Metabolism of Herbicides in Higher Plants,* CEPCO Division, Burgess Publishing Comnpany, Minneapolis, MN.
6. Cole, D.J., Edwards, R., and Owen, W.J. (1987), The role of metabolism in herbicide selectivity, in *Herbicides, Progress in Pesticide Biochemistry and Toxicology,* Vol. 6, D.H. Hutson and T.R. Roberts, Eds., John Wiley & Sons, Chichester, U.K., p. 57.
7. Kearney, P.C. and Kaufman, D.D., Eds. (1988), *Herbicides, Chemistry, Degradation and Mode of Action,* Vol. 3, Marcel Dekker, New York.
8. Hathway, D.E. (1989), *Molecular Mechanisms of Herbicide Selectivity,* Oxford University Press, New York.

9. Owen, W.J. (1989), Metabolism of herbicides — detoxification as a basis for selectivity, in *Herbicides and Plant Metabolism,* A.D. Dodge, Ed., Society for Experimental Biology Seminar Series 38, Cambridge University Press, London, p. 171.

10. Sandermann, H., Diesperger, H., and Scheel, D. (1977), Metabolism of xenobiotics by plant cell cultures, in *Plant Tissue Culture and Its Biotechnological Application,* W. Barz, E, Reinhard, and M.H. Zenk, Eds., Springer-Verlag, Berlin, p. 178.

11. Mumma, R.O. and Davidonis, G.H. (1983), Plant tissue culture and pesticide metabolism, in *Progress in Pesticide Biochemistry and Toxicology,* Vol. 3, D.H. Hutson and T.R. Roberts, Eds., John Wiley & Sons, Chichester, U.K., p. 255.

12. Sandermann, H., Scheel, D., and v.d. Trenk, T. (1984), Use of plant cell cultures to study the metabolism of environmental chemicals, *Ecotox. Environ. Saf.,* 8: 167.

13. Harms, H. and Langebartels, C. (1986), Standardized plant cell suspension test systems for an ecotoxicologic evaluation of the metabolic fate of xenobiotics, *Plant Sci.,* 45: 157.

14. Swisher, B.A. (1987), Use of plant cell cultures in pesticide metabolism studies, in *Biotechnology in Agricultural Chemistry,* ACS Symposium Series 334, H.M. LeBaron, R.O., Mumma, R.C., Honeycutt, and J.H. Duesing, Eds., American Chemical Society, Washington, D.C., p. 18.

15. Komoßa, D., Gennity, I., and Sandermann, H. (1992), Plant metabolism of herbicides with C-P bonds — Glyphosate, *Pestic. Biochem. Physiol.,* 43: 85.

16. Sandermann, H. (1992), Plant metabolism of xenobiotics, *TIBS,* 17: 82.

17. Sandermann, H. (1987), Pestizid-Rückstände in Nahrungspflanzen, Die Rolle des pflanzlichen Metabolismus, *Naturwissenschaften,* 74: 573.

18. Durst, F. (1991), Biochemistry and physiology of plant cytochrome P-450, in *Microbial and Plant Cytochromes P-450: Biochemical Characteristics, Genetic Engineering and Practical Implications,* K. Ruckpaul and H. Rein, Eds., Frontiers in Biotransformation, Vol. 4, Akademie Verlag, Berlin, p. 191.

19. Frear, D.S., Swanson, H.R., and Thalacker, F.W. (1991), Induced microsomal oxidation of diclofop, triasulfuron, chlorsulfuron, and linuron in wheat, *Pestic. Biochem. Physiol.,* 41: 274.

20. Fonné-Pfister, R., Gaudin, J., Kreuz, K., Ramsteiner, K., and Ebert, E. (1990), Hydroxylation of primisulfuron by an inducible cytochrome P-450-dependent monooxygenase system from maize, *Pestic. Biochem. Physiol.,* 37: 165.

21. Zimmerlin, A. and Durst, F. (1990), Xenbiotic metabolism in plants: aryl hydroxylation of diclofop by a cytochrome P-450 enzyme from wheast, *Phytochemistry,* 29: 1729.

22. McFadden, J.J., Frear, D.S., and Mansager, E.R. (1989), Aryl hydroxylation of diclofop by a cytochrome P-450-dependent monooxygenase from wheat, *Pestic. Biochem. Physiol.,* 34: 92.

23. Mougin, C., Cabanne, F., Canivenic, M.-C., and Scalla, R. (1990), Hydroxylation and *N*-demethylation of chlortoluron by wheat microsomal enzymes, *Plant Sci.,* 66: 195.

24. Mougin, C., Polge, N., Scalla, R., and Cabanne, F. (1991), Interactions of various agrochemicals with cytochrome P-450-dependent monooxygenases of wheat cells, *Pestic. Biochem. Physiol.,* 40: 1.

25. Fonné-Pfister, R. and Kreuz, K. (1990), Ring-methyl hydroxylation of chlortoluron by an inducible cytochrome P-450-dependent enzyme from maize, *Phytochemistry*, 29: 2793.

26. Burton, J.D., Moreland, D.E., and Corbin, F.T. (1990), *In vivo* and *in vitro* metabolism of bentazon, *Plant Physiol.*, 93 (Suppl.): 65.

27. Moreland, D.E., Corbin, F.T., Burton, J.D., and Maness, E.P. (1990), Metabolism of bentazon by excised tissue and a microsomal preparation from grain sorghum seedlings, in Abstr. 7th International Congress of Pesticide Chemistry, Hamburg, August 5–10, 1990, Vol. 2, H. Frehse, E. Kesseler-Schmitz, and S. Conway, Eds., Abstract No. 06E-05.

28. Moreland, D.E. and Corbin, F.T. (1991), Influence of safeners on the *in vivo* and *in vitro* metabolism of bentazon and metolachlor by grain sorghum shoots: a preliminary report, *Z. Naturforsch.*, 46c: 906.

29. McFadden, J.J., Gronwald, J.W., and Eberlein, C.V. (1990), *In vitro* hydroxylation of bentazon by microsomes from naphthalic anhydride-treated corn shoots, *Biochem. Biophys. Res. Commun.*, 168: 206.

30. Sandermann, H. (1982), Metabolism of environment chemicals: a comparison of plant and liver enzyme systems, in *Environmental Mutagenesis, Carcinogenesis and Plant Biology*, Vol. 1, E.J. Klekowski, Ed., Praeger Publishers, New York, p. 1.

31. Moreland, D.E., Corbin, F.T., and Novitzky, W.P. (1990), Metabolism of metolachlor by a microsomal fraction isolated from grain sorghum (*Sorghum bicolor*) shoots, *Z. Naturforsch.*, 45c: 558.

32. O'Keefe, D.P. and Leto, K.L. (1989), Cytochrome P-450 from the mesocarp of avocado (*Persea americana*), *Plant Physiol.*, 89: 1141.

33. Ryan, P.J., Gross, D., Owen, W.J., and Laanio, T.L. (1981), The metabolism of chlortoluron, diuron, and CGA 43 057 in tolerant and susceptible plants, *Pestic. Biochem. Physiol.*, 16: 213.

34. Owen, W.J. and Donzel, B. (1986), Oxidative degradation of chlortoluron, propiconazole, and metalaxyl in suspension cultures of various crop plants, *Pestic. Biochem. Physiol.*, 26: 75.

35. Cole, D.J. and Owen, W.J. (1987), Influence of monooxygenase inhibitors on the metabolism of the herbicides chlortoluron and metolachlor in cell suspension cultures, *Plant Sci.*, 50: 13.

36. Cole, D.J. and Owen, W.J. (1988), Metabolism of chlortoluron in cell suspensions of *Lactuca sativa*: a qualitative change with age of culture, *Phytochemistry*, 27: 1709.

37. Beyer Jr., E.M., Duffy, M.J., Hay, J.V., and Schlueter, D.D. (1988), Sulfonylureas, in *Herbicides, Chemistry, Degradation and Mode of Action*, Vol. 3, P.C. Kearney and D.D. Kaufman, Eds., Marcel Dekker, New York, p. 117.

38. Brown, H.M. and Kearney, P.C. (1991), Plant biochemistry, environmental properties, and global impact of the sulfonylurea herbicides, in *Synthesis and Chemistry of Agrochemicals*, Vol. 2, ACS Symposium Series 443, D.R. Baker, J.G. Fenyes, and W.K. Moberg, Eds., American Chemical Society, Washington, D.C., p. 32.

39. Sweetser, P.B., Schow, G.S., and Hutchison, J.M. (1982), Metabolism of chlorsulfuron by plants: biological basis for selectivity of a new herbicide for cereals, *Pestic. Biochem. Physiol.*, 17: 18.

40. Meyer, A.M. and Müller, F. (1989), Triasulfuron and its selective behaviour in wheat and *Lolium perenne,* in *Brighton Crop Protection Conference — Weeds,* Vol. 1, p. 441.

41. Blair, A.M. and Martin, T.D. (1988), A review of the activity, fate and mode of action of sulfonylurea herbicides, *Pestic. Sci.,* 22: 195.

42. Hatzios, K.K. and Hoagland, R.E., Eds., (1989), *Crop Safeners for Herbicides, Development, Uses, and Mechanisms of Action,* Academic Press, San Diego.

43. Hatzios, K.K. (1991), An overview of the mechanisms of action of herbicide safeners, *Z. Naturforsch.,* 46c: 819.

44. Ebert, E. and Kreuz, K. (1991), Die Selektivität von Herbiziden, Das Prinzip der Safener, *BiuZ,* 21: 298.

45. Clay, S.A. and Oelke, E.A. (1988), Basis for differential susceptibility of rice (*Oryza sativa*), wild rice (*Zizania palustris*), and giant burreed (*Sparganium eurycarpum*) to bentazon, *Weed Sci.,* 36: 301.

46. Sterling, T.M. and Balke, N.E. (1989), Differential bentazon metabolism and retention of bentazon metabolites by plant cell cultures, *Pestic. Biochem. Physiol.,* 34: 39.

47. Sterling, T.M. and Balke, N.E. (1990), Bentazon uptake and metabolism by cultured plant cells in the presence of monooxygenase inhibitors and cinnamic acid, *Pestic. Biochem. Physiol.,* 38: 66.

48. Gronwald, J.W. and Connelly, J.A. (1991), Effect of monooxygenase inhibitors on bentazon uptake and metabolism in maize cell suspension cultures, *Pestic. Biochem. Physiol.,* 40: 284.

49. Connelly, J.A., Johnson, M.D., Gronwald, J.W. and Wyse, D.L. (1988), Bentazon metabolism in tolerant and susceptible soybean (*Glycine max*) genotypes, *Weed Sci.,* 36: 417.

50. Sterling, T.M. and Balke, N.E. (1988), Use of soybean (*Glycine max*) and velvetleaf (*Abutilon theophrasti*) suspension-cultured cells to study bentazon metabolism, *Weed Sci.,* 36: 558.

51. Leah, J.M., Worrall, T.L., and Cobb, A.H. (1989), Metabolism of bentazone in soybean and the influence of tetcyclasis, BAS 110 and BAS 111, *Brighton Crop Protection Conference — Weeds,* Vol. 1, p. 433.

52. Leah, J.M., Worrall, T.L., and Cobb, A.H. (1991), A study of bentazon uptake and metabolism in the presence and the absence of cytochrome P-450 and acetylcoenzyme A carboxylase inhibitors, *Pestic. Biochem. Physiol.,* 39: 232.

53. Yaacoby, T., Hall, J.C., and Stephenson, G.R. (1991), Influence of fenchloazole-ethyl on the metabolism of fenoxaprop-ethyl in wheat, barley, and crabgrass, *Pestic. Biochem. Physiol.,* 41: 296.

54. Koeppe, M.K., Anderson, J.J., and Shalaby, L.M. (1990), Metabolism of [^{14}C]-quizalofop-ethyl in soybean and cotton plants, *J. Agric. Food Chem.,* 38: 1085.

55. Kreuz, K., Gaudin, J., Stingelin, J., and Ebert, E. (1991), Metabolism of the aryloxyphenoxypropanoate herbicide, CGA 184927, in wheat, barley and maize: Differential effects of the safener, CGA 185072, *Z. Naturforsch.,* 46c: 901.

56. Frear, D.S., Swanson, H.R., and Mansager, E.R. (1983), Acifluorfen metabolism in soybean: diphenylether bond cleavage and the formation of homoglutathione, cysteine and glucose conjugates, *Pestic. Biochem. Physiol.,* 20: 299.

57. Lamoureux, G.L., Rusness, D.G., Schröder, P., and Rennenberg, H. (1991), Diphenyl ether herbicide metabolism in a spruce cell suspension culture: the identification of two novel metabolites derived from a glutathione conjugate, *Pestic. Biochem. Physiol.,* 39: 291.

58. Weimer, M.R., Balke, N.E., and Buhler, D.D. (1992), Herbicide clomazone does not inhibit *in vitro* geranylgeranyl synthesis from mevalonate, *Plant Physiol.,* 98: 427.

59. Weimer, M.R., Balke, N.E., and Buhler, D.D. (1992), Absorption and metabolism of clomazone by suspension-cultured cells of soybean and velvetleaf, *Pestic. Biochem. Physiol.,* 42: 43.

60. Roberts, T.R. (1981), The metabolism of the synthetic pyrethroids in plants and soils, in *Progress in Pesticide Biochemistry,* Vol. 1, D.H. Hutson and T.R. Roberts, Eds., John Wiley & Sons, Chichester, U.K., p. 115.

61. Vonk, J.W. (1983), Metabolism of fungicides in plants, in *Progress in Pesticide Biochemistry and Toxicology,* Vol. 3, D.H. Hutson and T.R. Roberts, Eds., John Wiley & Sons, Chichester, U.K., p. 111.

62. Edwards, V.T. and McMinn, A.L. (1985), Biotransformation of pesticides and other xenobiotics in plants and soils — recent developments, in *Progress in Pesticide Biochemistry and Toxicology,* Vol. 4, D.H. Hutson and T.R. Roberts, Eds., John Wiley & Sons, Chichester, U.K., p. 103.

63. Arjmand, M. and Sandermann, H. (1985), Metabolism of DDT and related compounds in cell suspension cultures of soybean (*Glycine max* L.) and wheat (*Triticum aestivum* L.), *Pestic. Biochem. Physiol.,* 23: 389.

64. Krell, H.-W. and Sandermann, H. (1985), Compartmentation of the plant metabolic enzymes for the persistent plasticizer chemical, bis-2(-ethylhexyl)-phthalate, *Plant Sci.,* 40: 87.

65. Krell, H.-W. and Sandermann, H. (1986), Metabolism of the persistent plasticizer chemical bis(2-ethylhexyl)phthalate in cell suspension cultures of wheat (*Triticum aestivum* L.). Discrepancy from the intact plant, *J. Agric. Food Chem.,* 34: 194.

66. Lee, I. and Fletcher, J.S. (1990), Metabolism of polychlorinated biphenyls (PCBS) by plant tissue cultures, in *Progress in Plant Cellular and Molecular Biology,* Proceedings of the 7th International Congress on Plant Tissue and Cell Culture, Amsterdam, June 24–29, 1990, H.J.J. Nijkamp, L.H. W. v.d. Plas, and J. Aatrijk, Eds., Kluwer Academic Publishers, Dordrecht, The Netherlands, p. 656.

67. Lee, I. and Fletcher, J.S. (1992), Involvement of mixed function oxidase systems in polychlorinated biphenyl metabolism by plant cells, *Plant Cell Rep.,* 11: 97.

68. Schäfer, W. and Sandermann, H. (1988), Metabolism of pentachlorophenol in cell suspension cultures of wheat (*Triticum aestivum* L.). Tetrachlorocatechol as a primary metabolite, *J. Agric. Food Chem.,* 36: 370.

69. Schmitt, R., Kaul, J., v.d. Trenck, T., Schaller, E., and Sandermann, H. (1985), β-D-Glucosyl and *O*-malonyl-β-D-glucosyl conjugates of pentachlorophenol in soybean and wheat: identification and enzymatic synthesis, *Pestic. Biochem. Physiol.,* 24: 77.

70. Duke, S.O. and Kenyon, W.H. (1988), Polycyclic alkanoic acids, in *Herbicides, Chemistry, Degradation and Mode of Action,* Vol. 3, P.C. Kearney and D.D. Kaufmann, Eds., Marcel Dekker, New York, p. 71.

71. Shimabukuro, R.H. (1985), Detoxication of herbicides, in *Weed Physiology,* Vol. 2, Herbicide Physiology, S.O. Duke, Ed., CRC Press, Boca Raton, FL, p. 215.

72. Shimabukuro, R.H. and Hoffer, B.L. (1991), Metabolism of diclofop-methyl in susceptible and resistant biotypes of *Lolium rigidum, Pestic. Biochem. Physiol.,* 39: 251.

73. Holtum, J.A.M., Matthews, J.M., Häusler, R.E., Liljegren, D.R., and Powles, S.B. (1991), Cross-resistance to herbicides in annual ryegrass (*Lolium rigidum*). III. On the mechanism of resistance to diclofop-methyl, *Plant Physiol.,* 97: 1026.

74. Devine, M.D., MacIsaac , S.A., Romano, M.L., and Hall, J.C. (1992), Investigation of the mechanism of diclofop resistance in two biotypes of *Avena fatua, Pestic. Biochem. Physiol.,* 42: 88.

75. Tanaka, F.S., Hoffer, B.L., Shimabukuro, R.H., Wien, R.G., and Walsh, W.C. (1990), Identification of the isomeric hydroxylated metabolites of methyl 2-[4-(2,4-dichlorophenoxy)phenoxy]propanoate (diclofop-methyl) in wheat, *J. Agric. Food Chem.,* 38: 559.

76. Dahroug, S. and Müller, F. (1990), Response of susceptible and tolerant plants to diclofop-methyl, *J. Plant Dis. Prot.,* 97, 154.

77. Brown, H.M., Wittenbach, V.A., Forney, D.R., and Strachan, S.D. (1990), Basis for soybean tolerance to thifensulfuron methyl, *Pestic. Biochem. Physiol.,* 37: 303.

78. Brown, H.M., and Neighbors, S.M. (1987), Soybean metabolism of chlorimuron ethyl: physiological basis for soybean selectivity, *Pestic. Biochem. Physiol.,* 29: 112.

79. Schulz, A., Wengenmayer, F., and Goodman, H.M. (1990), Genetic engineering of herbicide resistance in higher plants, *Crit. Rev. Plant Sci.,* 9: 1.

80. Steinrücken, H.C. and Amrhein, N. (1980), The herbicide glyphosate is a potent inhibitor of 5-enolpyruvylshikimic acid-3-phosphate synthase, *Biochem. Biophys. Res. Commun.,* 94: 1207.

81. Köcher, H. (1989), Inhibitors of glutamine synthetase and their effects in plants, in *Prospects for Amino Acid Biosynthesis Inhibitors in Crop Protection and Pharmaceutical Chemistry,* BCPC Monograph No. 42, British Crop Protection Council, p. 173.

82. Mazur, B.J. and Falco, S.C. (1989), The development of herbicide resistant crops, *Annu. Rev. Plant Physiol. Plant Mol. Biol.,* 40: 441.

83. Dekeyser, R., Inzé, D., and v. Montagu, M. (1990), Transgenic plants, in *Gene Manipulation in Plant Improvement,* Vol. 2, 19th Stadler Genetics Symposium, J.P. Gustafson, Ed., Plenum Press, New York, p. 237.

84. Komoßa, D. and Sandermann, H. (1992), Plant metabolism of herbicides with C-P bonds — Phosphinothricin, *Pestic. Biochem. Physiol.,* 43: 95.

85. Lamoureux, G.L. and Rusness, D.G. (1986), Xenobiotic conjugation in higher plants, in *Xenobiotic Conjugation Chemistry,* ACS Symposium Series 299, G.L. Gilson, J. Caldwell, D.H. Hutson, and J.J. Menn, Eds., American Chemical Society, Washington, D.C., p. 62.

86. Lamoureux, G.L. Shimabukuro, R.H., and Frear, D.S. (1991), Glutathione and glucoside conjugatgion in herbicide selectivity, in *Herbicide Resistance in Weeds and Crops,* J.C. Caseley, G.W. Cussans, and R.K. Atkin, Eds., 11th Long Ashton International Symposium (Bristol 1989), Butterworth, Oxford, p. 227.

87. Sandermann, H., Schmitt, R., Eckey, H., and Bauknecht, T. (1991), Plant biochemistry of xenobiotics: isolation and properties of soybean *O*- and *N*-glucosyl and *O*- and *N*-malonyltransferases for chlorinated phenols and anilines, *Arch. Biochem. Biophys.,* 287: 341.

88. Leah, J.M., Worrall, T.L., and Cobb, A.H. (1992), Isolation and characterization of two glucosyltransferases from *Glycine max* associated with bentazone metabolism, *Pestic. Sci.,* 34: 81.

89. Hatzios, K.K. and Penner, D. (1988), Metribuzin, in *Herbicides, Chemistry, Degradation and Mode of Action,* Vol. 3, P.C. Kearney and D.D. Kaufman, Eds., Marcel Dekker, New York, p. 191.

90. Davis, D.G., Olson, P.A., Swanson, H.R., and Frear, D.S. (1991), Metabolism of the herbicide metribuzin by an *N*-glucosyltransferase from tomato cell cultures, *Plant Sci.,* 74: 73.

91. Frear, D.S. (1968), Herbicide metabolism in plants. I. Purification and properties of UDP-glucoside:arylamine *N*-glucosyl-transferase from soybean, *Phytochemistry,* 7: 381.

92. Metschulat, G. (1990), Untersuchungen zum Xenobiotika-Stoffwechsel in pflanzlichen zellsuspensionskulturen, Doctoral thesis, University of Münster, Germany.

93. Jordan, M. (1990), Untersuchungen zum Metabolismus eines Insektizids in Zellkulturen von Chenopodium rubrum, Doctoral thesis, University of Münster, Germany.

94. Barz, W., Jordan, M., and Metschulat, G. (1990), Bioconversion of xenobiotics (pesticides) in plant cell cultures, in *Progress in Plant Cellular and Molecular Biology,* Proceedings of the 7th International Congress on Plant Tissue and Cell Culture, Amsterdam, June 24–29, 1990, H.J.J. Nijkamp, L.H.W. v.d. Plas, and J. Aartrijk, Eds., Kluwer Academic Publishers, Dordrecht, The Netherlands, p. 631.

95. Harborne, J.B., Ed. (1988), *The Flavonoids, Advances in Research since 1980,* Chapman & Hall, London.

96. Giese, M. (1992), Untersuchungen zum Metabolismus von 1-*N*(malonylamino)-cyclopropan-1-carbonsäure in *Pisum sativum* L., Doctoral thesis, University of Halle, Germany.

97. Höhl, U. (1988), Untersuchungen zum Metabolismus eines Insektizids in Pflanzen und Zellkulturen der Sojabohne, Doctoral thesis, University of Münster, Germany.

98. Lamoureux, G.L. and Rusness, D.G. (1989), The role of glutathione and glutathione-*S*-transferases in pesticide metabolism, selectivity, and mode of action in plants and insects, in *Glutathione,* Part B, Chemical, Biochemical, and Medical Aspects, D. Dolphen, O. Avramovic, and R. Poulson, Eds., John Wiley & Sons, New York, p. 153.

99. Lamoureux, G.L. and Rusness, D.G. (1989), Propachlor metabolism in soybean plants, excised soybean tissues, and soil, *Pestic. Biochem. Physiol.,* 34: 187.

100. Han, S. and Hatzios, K.K. (1991), Uptake, translocation, and metabolism of [^{14}C]pretilachlor in fenclorim-safened and unsafened rice seedlings, *Pestic. Biochem. Physiol.,* 39: 281.

101. Fuerst, E.P. and Lamoureux, G.L. (1992), Mode of action of the dichloroacetamide antidote BAS 145-138 in corn. II. Effects on metabolism, absorption, and mobility of metazachlor, *Pestic. Biochem. Physiol.,* 42: 78.

102. Breaux, E.J. (1986), Identification of the initial metabolites of acetochlor in corn and soybean seedlings, *J. Agric. Food Chem.,* 34: 884.

103. LeBaron, H.M., McFarland, J.E., and Simoneaux, B.J. (1988), Metolachlor, in *Herbicides, Chemistry, Degradation and Mode of Action*, Vol. 3, P.C. Kearney and D.D. Kaufman, Eds., Marcel Dekker, New York, p. 335.
104. Lamoureux, G.L. and Rusness, D.G. (1987), EPTC metabolism in corn, cotton, and soybean: Identification of a novel metabolite derived from the metabolism of a glutathione conjugate, *J. Agric. Food Chem.*, 35: 1.
105. Lamoureux, G.L. and Rusness, D.G. (1986), Tridiphane [2-(3,5-dichlorophenyl)-2-(2,2,2-trichloroethyl)oxirane], an atrazine synergist: enzymatic conversion to a potent glutathione S-transferase inhibitor, *Pestic. Biochem. Physiol.*, 26: 323.
106. Lamoureux, G.L., Fuerst, E.P., and Rusness, D.G. (1990), Selected aspects of glutathione conjugation research in herbicide metabolism and selectivity, in *Sulfur Nutrition and Sulfur Assimilation in Higher Plants, Fundamental Environmental and Agricultural Aspects*, H. Rennenberg, C. Brunold, L.J. deKok, and I. Stulen, Eds., SPB Academic Publishing, The Hague, p. 217.
107. Timmermann, K.P. (1989), Molecular characterization of corn glutathione S-transferase isozymes involved in herbicide detoxication, *Physiol. Plant.*, 77: 465.
108. Langebartels, C. and Harms, H. (1985), Analysis for nonextractable (bound) residues of pentachlorophenol in plant cells using a cell wall fractionation procedure, *Ecotox. Environ. Saf.*, 10: 268.
109. Pogány, E., Pawlitzki, K.-H., and Wallnöfer, P.R. (1990), Formation, distribution and bioavailability of cell wall bound residues of 4-chloroaniline and 2,4-dichlorophenol, *Chemosphere*, 21: 349.
110. Sandermann, H., Musick, T.J., and Aschbacher, P.W. (1992), Animal bioavailability of a 3,4-dichloroaniline-lignin metabolite fraction from wheat, *J. Agric. Food Chem.*, 40: 2001.
111. Sandermann, H., Scheel, D., and v.d. Trenck, T. (1983), Metabolism of environmental chemicals by plants — copolymerization into lignin, *J. Appl. Polym. Sci. Appl. Polym. Symp.*, 37: 407.
112. Khan, S.U. and Dupont, S. (1987), Bound pesticide residues and their bioavailability, in *Pesticide Science and Technology*, R. Greenhalgh and T.R. Roberts, Eds., Blackwell Scientific, Oxford, p. 417.
113. Harms, H. (1992), *In-vitro* systems for studying phytotoxicity and metabolic fate of pesticides and xenobiotics in plants, *Pestic. Sci.*, 35: 277.
114. Schmitt, R. and Sandermann, H. (1982), Specific localization of β-D-glucoside conjugates of 2,4-dichlorophenoxyacetic acid in soybean vacuoles, *Z. Naturforsch.*, 37c: 772.
115. Kreis, K. and Hölz, H. (1991), Zellulärer Transport und Speicherung von Naturstoffen. Die Bedeutung der Vakuole im pflanzlichen Stoffwechsel, *Naturwissenschaften*, 44: 463.
116. Coupland, D. (1991), The role of compartmentation of herbicides and their metabolites in resistance mechanisms, in *Herbicide Resistance in Weeds and Crops*, J.C. Caseley, G.W. Cussans, and R.K. Atkin, Eds., 11th Long Ashton International Symposium (Bristol 1989), Butterworth-Heinemann, Oxford, p. 263.
117. Metcalf, R.L., Sangha, G.K., and Kapoor, I.P. (1971), Model ecosystem for the evaluation of pesticide biodegradability and ecological magnification, *Environ. Sci. Technol.*, 5: 709.

118. Cole, L.K., Metcalf, R.L., and Sanborn, J.R. (1976), Environmental fate of insecticides in terrestrial model ecosystems, *Int. J. Environ. Stud,.* 10: 7.

119. Führ, F. (1985), Application of ^{14}C-labeled herbicides in lysimeter studies, *Weed Sci.,* 33: 11.

120. Kubiak, R., Führ, F., and Mittelstaedt, W. (1990), Comparative studies on the formation of bound residues in soil in outdoor and laboratory experiments, *Int. J. Environ. Anal. Chem.,* 39: 47.

121. Schuphan, I., Schärer, E., Heise, M., and Ebing, W. (1987), Use of laboratory model ecosystems to evaluate quantitatively the behaviour of chemicals, in *Pesticide Science and Biotechnology,* R. Greenhalgh and T.R. Roberts, Eds., Blackwell Scientific, Oxford, p. 437.

122. Mc Farlane, C., Pfleeger, T., and Fletcher, J. (1990), Effect, uptake and disposition of nitrobenzene in several terrestrial plants, *Environ. Toxicol. Chem.,* 9: 513.

123. Scheunert, I., Schroll, R., and Trapp, S. (1993), Prediction of plant uptake of organic xenobiotics by chemical substance and plant properties, in *Chemical Exposure Predictions,* D. Calamari, Ed., Lewis Publishers, Chelsea, MI.

124. Schroll, R. and Scheunert, I. (1992), A laboratory system to determine separately the uptake of organic chemicals from soil by plant roots and by leaves after vaporization, *Chemosphere,* 24: 97.

125. Trapp, S., Matthies, M., Scheunert, I., and Topp, E.M. (1990), Modeling the bioconcentration of organic chemicals in plants, *Environ. Sci. Technol.,* 24: 1246.

126. Mumma, R.O. and Hamilton, R.H. (1979), Xenobiotic metabolism in higher plants: *In vitro* tissue and cell culture techniques, in *Xenobiotic Metabolism: In Vitro Methods,* ACS Symposium Series 97, G.D. Paulson, D.S. Frear, and E.P. Marks, Eds., American Chemical Society, Washington, D.C., p. 35.

127. Wickliff, C. and Fletcher, J.S. (1991), Tissue culture as a method for evaluating the biotransformation of xenobiotics by plants, in *Plants for Toxicity Assessment,* Vol. 2, J.W. Gorsuch, W.R. Lower, W. Wang, and M.A. Lewis, Eds., American Society for Testing and Materials, Philadelphia, p. 250.

128. Gressel, J. (1987), *In vitro* plant cultures for herbicide prescreening, in *Biotechnology in Agricultural Chemistry,* ACS Symposium Series 334, H.M. LeBaron, R.O. Mumma, R.C. Honeycutt, and J.H. Duesing, Eds., American Chemical Society, Washington, D.C., p. 41.

129. Grossman, K. (1988), Plant cell suspensions for screening and studying the mode of action of plant growth retardants, *Adv. Cell Culture,* 6: 89.

130. Ebing, W., Haque, A., Schuphan, I., Harms, H., Langebartels, C., Scheel, D., v.d. Trenck, K.T., and Sandermann, H. (1984), Ecochemical assessment of environmental chemicals: draft guideline of the test procedure to evaluate metabolism and degradation of chemicals by plant cell cultures, *Chemosphere,* 13: 947.

131. Wolkening, A., Schuphan, I., Ebing, W., Lundehn, J.-R., Nolting, H.-G., Parnemann, H., and Röpsch, A. (1988), Richtlinienentwurf zum Verbleib von Pflanzenschutzmitteln in und auf Pflanzen — Abbau und Metabolismus, *Mitt. Biol. Bundesanst. Land Forstwirtsch. Berlin-Dahlem,* 245: 406.

132. Gamborg, O.L., Miller, R.A., and Ojima, K. (1968), Nutrient requirements of suspension cultures of soybean root cells, *Exp. Cell Res.,* 50: 151.

133. Harms, H. (1973), Pflanzliche Zellsuspensionskulturen — ihr Leistungsvermögen für Stoffwechseluntersuchungen, *Landbauforsch. Völkenrode,* 23: 127.

134. Schmidt, B., Ebing, W., and Schuphan, I. (1988), Einsatz eines Pflanzenzellkultur-Tests zur Ermittlung der Metabolisierbarkeit von Pflanzenschutzmitteln, *Gesunde Pflanzen,* 40: 245.
135. Winkler, R. and Sandermann, H. (1989), Plant metabolism of chlorinated anilines: isolation and identification of *N*-glucosyl and *N*-malonyl conjugates, *Pestic. Biochem. Physiol.,* 33: 239.
136. Industrieverband Agrar, Ed. (1990), Wirkstoffe in Pflanzenschutz- und Schädlingsbekämpfungsmitteln: Physikalisch-chemische und toxikologische Daten, BLV, Munich, Germany.
137. Haque, A., Ebing, W., and Schuphan, I. (1984), Ecochemical assessment of environmental chemicals. II. Standard supplementary screening procedure to evaluate more volatile and less persistent chemicals in plant cell cultures, *Chemosphere,* 13: 315.
138. Langebartels, C. and Harms, H. (1986), Plant cell suspension cultures as test systems for an ecotoxicologic evaluation of chemicals. Growth inhibition effects and comparison with the metabolic fate in intact plants, *Angew. Bot.,* 60: 113.
139. Schuphan, I., Schmidt, B., and Veit, P. (1990), Pesticide metabolism and degradation in plants, *Gesunde Pflanzen,* 42: 276.
140. Krell, H.-W. and Sandermann, H. (1984), Plant biochemistry of xenobiotics. Purification and properties of a wheat esterase hydrolyzing the plasticizer chemical, bis(2-ethylhexyl)phthalate, *Eur. J. Biochem.,* 143: 57.
141. Raynaud, S., Bastide, J., and Coste, C. (1985), Use of a mathematical model to determine the fate of atrazine in barley (*Hordeum vulgare*) plants, *Weed Sci.,* 33: 906.

Part Three
Modeling

CHAPTER 5

Model for Uptake of Xenobiotics into Plants

Stefan Trapp

TABLE OF CONTENTS

1-56670-078-7/95/$0.00+$.50
© 1995 by CRC Press, Inc.

I. INTRODUCTION

The modern worldwide chemical industry produces thousands of substances in large amounts. Most of them were not present in the biosphere before (i.e., they are xenobiotic), and some of them are known to be toxic to organisms. During or at the end of the life cycle of industrial products, chemicals are released into air, water, and/or soil. From there, they probably find their way into living organisms.

Terrestrial plants are contaminated in various ways with xenobiotic substances. The hazard is especially large when plants are growing on polluted sites. This includes waste sites and areas where contaminated sewage sludge has been applied. Another risk is the uptake of pollutants from air, because plants are particularily exposed toward an uptake of airborne substances. However, uptake of toxic compounds is not only dangerous to plants. At the same time this is the first step in the contamination of the food web and of man, too.

On the other hand, there are cases where the uptake of toxic substances is necessary for pest control: systemic herbicides and other systemic pesticides can only unfold their action when they reach the target within the plants. Tools are needed for the interpretation and prediction of the fate of chemicals in plants. One major chance herefore are mathematical models that combine the physicochemical properties of chemicals with the anatomical and physiological circumstances of plants. In this chapter, such a model* is developed, tested, and applied. The goal is the dynamic calculation of transport, partitioning, and metabolism of anthropogenic organic chemicals in roots, stem, leaves, and fruits of plants.

For the mathematical description of the numerous processes within a living plant, the reduction to relevant processes and an abstraction is necessary. This is why some simplifying assumptions have to be made. A test with a number of laboratory experiments can prove the validity of the model.

II. CHEMODYNAMIC CONCEPTS IN POLLUTANT BEHAVIOR

The work requires an interdisciplinary view. This means that knowledge of biology, chemistry, physics, mathematics, and soil science is needed. Every division has certain defined expressions. Transport and partitioning of chemicals in the environment may be described with the concepts of chemodynamics.[1] It follows an introduction to basic principles. This may be a help for understanding the mathematical description of chemical behavior in plants. Symbols are listed in Table 1.

A. EQUILIBRIUM PARTITIONING BETWEEN NONMIXABLE PHASES

If a substance is dissolved until it is soluble in two adjacent, nonmixable phases, the ratio of the concentrations in these two phases will have a certain value. At this value, the *fugacities* of the substance are the same in both phases. Equal fugacities will also appear at concentrations below maximum solubility, when the substance is completely mixed in both phases. Then the system is in equilibrium. As long as concentrations are small (approximately below 10% of the solubility), the fugacity is directly proportional to the concentration of a

*The plant exposure model PlantX can be ordered on request without charge from the author.

substance.[2] In this case, the equilibrium is independent from the absolute concentration. This is a basic assumption to the following considerations. The *partition coefficient,* K, describes the equilibrium concentration ratio:

$$K_{ij} = C_i/C_j \qquad (2.1)$$

where K is the partition coefficient (kg/m^3 to kg/m^3 or mol/m^3 to mol/m^3), C is the equilibrium concentration (kg/m^3 or mol/m^3), and i and j are the indices of the adjacent phases.

In the environment, different phases exist, too. The partitioning of a liquid between air and water is described by the Henry's law constant, H. It can be calculated from solubility in water and saturation vapor pressure in air:[3]

$$H = P_s/S \qquad (2.2)$$

where H is the Henry's law constant ($Pa \cdot m^3/mol$), P_s is the saturation vapor pressure (Pa; at solids: of the subcooled liquid), and S is the aqueous solubility of the substance (mol/m^3). The atmosphere/water partition coefficient, K_{AW} (also named the dimensionless Henry's law constant), follows:

$$K_{AW} = H/(R \cdot T) = C_A/C_W \qquad (2.3)$$

where C_A is the equilibrium concentration in the atmosphere (kg/m^3), C_W is the equilibrium concentration in water (kg/m^3), R is the universal gas constant ($8.314 \, J \cdot mol^{-1} \cdot K^{-1}$), and T is the temperature (Kelvin). Henry's law constants may differ over many orders of amounts (for examples see Chapter 3 or Reference 4).

The equilibrium partitioning between a hydrophobic phase and water is often described by the 1-octanol/water partition coefficient, K_{OW} (ml/g). For many compounds this value was measured or estimated (see, for instance, References 3 and 5):

$$K_{OW} = C_O/C_W \qquad (2.4)$$

C_O is the equilibrium concentration of a substance in octanol, and C_W is the equilibrium concentration in water. The K_{OW} is used as a predictor for the partitioning between lipid phases in the environment (e.g., fish fat) and water.[6] The K_{OW} can vary over some orders of magnitude, too (for examples see Chapter 3).

The adsorption to solids describes the empirical Freundlich equation:[1]

$$x/m_M = K \cdot C_W^{1/n} \qquad (2.5)$$

where x is the amount of chemical adsorbed (g), m_M is the mass of sorbent M (g), K is the proportionality factor (Freundlich constant; cm^3/g), C_W is the equilibrium concentration in the aqueous phase (g/cm^3), and n is a measure for the nonlinearity of the relation. For small concentrations the values of n are close to one;[1] then the Freundlich constant may be seen as the slope of the linear adsorption/desorption isotherm. This method of describing sorption is usually the result of empirical studies, e.g., the root concentration factor.

If the sorbent is the soil matrix, the Freundlich constant is equal to the distribution coefficient K_d between soil matrix and water (cm^3/g):

$$x/m_M = C_M = K_d \cdot C_W \qquad (2.6)$$

where C_M is the concentration of chemical sorbed to the soil matrix (g/g). The natural bulk soil consists of soil matrix, soil water, and soil air. The bulk soil/water partition coefficient, K_{BW} is subsequently

$$K_{BW} = C_B/C_W = \rho_B \cdot K_d + \theta + (\varepsilon - \theta) \cdot K_{AW} \qquad (2.7)$$

where C_B is the equilibrium concentration in the bulk soil (kg/m^3), θ is the volumetric water fraction of the soil, ε is the volumetric total porosity of the soil, and ρ_B is the bulk density (dry, g/cm^3). The sorption of hydrophobic organic chemicals to the soil matrix is proportional to the organic content:[7]

$$K_d = K_{OC} \cdot OC \qquad (2.8)$$

where K_{OC} is the partition coefficient between organic carbon and water and OC is the organic carbon content of the soil (g/g dry mass). The K_{OC} is closely correlated with the K_{OW}, and some regressions have been published (see, e.g., Reference 3), among them

$$K_{OC} = 0.411 \cdot K_{OW} \text{ (Reference 7)} \qquad (2.9)$$

$$\log K_{OC} = 0.72 \cdot \log K_{OW} + 0.49 \text{ (Reference 8)} \qquad (2.10)$$

If partition constants with the same unit are used (here: kg/m^3 to kg/m^3), then it's possible to calculate an unknown value from two known ones:

$$K_{il} = K_{ij}/K_{lj} \qquad (2.11)$$

Air, water, and hydrophobic phases may be found everywhere in the environment. The calculation of equilibrium partition coefficients enables the estimation of the tendency of chemicals to partition. The equilibrium is the condition with the highest entropy. This condition is theoretically not reached within limited time periods. However, diffusion always occurs in the direction of growing entropy — i.e., in the direction of the equilibrium. The smaller the scale, the earlier "quasi"-equilibrium is achieved (local equilibrium).

B. TRANSPORT PROCESSES

Equilibrium is achieved only when complete mixing has occurred. Mixing requires a flux within the phases and across the boundary. In this chapter passive transport mechanisms based on diffusion and on mass flow with the surrounding medium are considered.

1. Diffusion

The microscopic movement of molecules results in a mixing and a macroscopic compensation of concentration differences. This relation was formulated mathematically by Fick in 1855. The first Fick's law of diffusion describes the net flux across a unit area in an isotropic medium by a constant concentration gradient:[9]

$$J = -D \cdot \partial C/\partial x \qquad (2.12)$$

J is the net flux of substance per unit area (kg $\cdot$ s^{-1} $\cdot$ m^{-2}), D is the diffusion coefficient (m^2/s), and $\partial C/\partial x$ is the concentration gradient (kg $\cdot$ m^{-3} $\cdot$ m^{-1}). If ∂x is small ($\Delta x \rightarrow 0$), $\partial C/\partial x$ can be replaced by $\Delta C/\Delta x$. The net flux N of the substance (kg/s) is directly proportional to the area A (m^2):

$$N = J \cdot A = -A \cdot D \cdot \Delta C/\Delta x \qquad (2.13)$$

The ratio between diffusion coefficient and pathway gives the conductance g (m/s). This is a measure of the velocity of the exchange. The reciprocal of g is the resistance r (s/m):[10]

$$D/\Delta x = 1/r = g \qquad (2.14)$$

The diffusion coefficient D for gases is around 10^{-5} to 10^{-4} m2/s, for liquids ca. 10^{-9} m^2/s, and for solids ca. 10^{-14} m^2/s. It can be seen that D primarily depends on the aggregate state and that diffusion within gases is most important.[9] Furthermore, D is proportional to the temperature and indirectly proportional to the molecular volume. Therefore, D may be estimated via the square root of molecular weights.[1] The ratio of the diffusion coefficients of two chemicals in the same medium is approximately

$$D_i/D_j = (MW_j/MW_i)^{0.5} \qquad (2.15)$$

where MW is the molecular weight (g/mol) and i and j are indices for two different chemicals. If more than one resistance has to be considered, the total resistance is calculated by the same equations as used for the electrical resistance.[10]

Diffusion across phase boundaries requires the consideration of partitioning. C_i and C_j are the concentrations in phases i and j, respectively, g is the conductance for the diffusive exchange, and K_{ij} is the partition coefficient. In equilibrium Equation 2.1 holds:

$$C_j/C_i = K_{ji} \text{ rsp. } C_i = C_j/K_{ji}$$

The gradient $\Delta C_{ij}/\Delta x$ (in the case of nonequilibrium) is

$$\Delta C_{ij}/\Delta x = (C_i - C_j/K_{ji})/\Delta x \qquad (2.16)$$

Then the flux between phases i and j, N_{ij} (kg/s), is

$$N_{ij} = -A \cdot g \cdot (C_i - C_j/K_{ji}) = -N_{ji} \qquad (2.17)$$

The net diffusion between the phases exists as long as C_i is unequal to C_j/K_{ji}.

2. Mass Flux

The flux N (kg/s) due to advection of the medium may be described mathematically by the equation

$$N = Q \cdot C \qquad (2.18)$$

where Q is the mass flux of the flowing medium (m³/s) due to the advection and C is the concentration of the chemical within the flowing medium (kg/m³).

The combination of the diffusive transport term with the advective transport term yields the diffusion-advection equation. Diffusion and advection are universal processes. Therefore, their description is the basis for many transport models of chemicals in water (e.g., Reference 11), soil (e.g., Reference 12), and air (e.g., Reference 13). In plants, chemicals are passively transported by advection and diffusion too.

C. THE FUGACITY CONCEPT

With fugacities, the same processes can be described in a different manner.[2,14] The basic equation of the fugacity approach is

$$C = f \cdot Z \qquad (2.19)$$

where C is the concentration (mol/m³), f is the fugacity (Pa), and Z is the fugacity capacity (mol · m⁻³ · Pa⁻¹). Diffusive transport between phases can be calculated elegantly and vividly because in equilibrium the fugacities between phases are equal ($f_i = f_j$). It follows for the equilibrium

$$C_i/C_j = (f_i \cdot Z_i)/(f_j \cdot Z_j) = Z_i/Z_j = K_{ij} \qquad (2.20)$$

Partition coefficients between material X and water W can be recalculated to give Z values by dividing through the Henry's law constant, H (Pa $\cdot$ m^3/mol), because $Z_W = 1/H$:

$$Z_X = K_{XW} \cdot Z_W = K_{XW}/H \qquad (2.21)$$

For every phase a Z value must be derived. The combination of Z values gives partition coefficients for any combination of phases. From the capacities Z the potential for accumulation can be easily seen.

Maybe the fugacity concept is somewhat confusing for readers not familiar with it. In the two following model approaches the fugacities will be used. In this contribution, the more "traditional" approach of partition constants to water is kept up. Note that the difference is only the formulation.

D. METABOLISM AND DEGRADATION

Xenobiotic chemicals that are located within plant cells are in a living and, therefore, very reactive environment (see Chapter 4). The time course of a reaction is described by the reaction kinetics. A reaction follows first-order kinetics when the rate is directly proportional to the concentration of the chemical present:

$$dC/dt = -\lambda \cdot C \qquad (2.22)$$

where λ is the reaction rate first order (1/time). If the concentration of one partner exceeds that of the other, then the reaction is often first order.

The half-life, $t_{1/2}$, is related to λ by the equation

$$t_{1/2} = \ln 2/\lambda \qquad (2.23)$$

Reactions are termed second order when the reaction velocity is proportional to two partners, k and l:

$$dC_k/dt = -\lambda' \cdot C_k \cdot C_l \qquad (2.24)$$

Here, λ' is the reaction rate second order with units 1/(time $\cdot$ concentration). Second-order reactions may occur when concentrations of two reaction partners are approximately the same. Reactions of higher order are rare.[15]

III. MODEL DEVELOPMENT

Based on the knowledge of the anatomical and physiological features of plants and the significant chemical properties and processes described earlier in this book, a model for the uptake and fate of anthropogenic organic compounds is deduced. This is done step by step. First, the single processes are formulated. Together they form the model for the whole plant. The model describes the dynamic uptake, metabolism, and accumulation of anthropogenic organic chemicals in roots, stem, leaves, and fruits. It is applicable to different plant species and most (nondissociating) organic chemicals.

The following processes are considered (Figure 1):

- Diffusive exchange between soil and roots in water and air pores
- Transfer into roots with the transpiration stream
- Translocation in the plant with the transpiration stream
- Partitioning into the stem
- Transport with the assimilation stream
- Diffusive exchange between air and leaf via stomata and cuticle
- Metabolism
- Dilution by growth

A. BASIC ASSUMPTIONS

The model is based on two assumptions that simplify model development and thereby make it possible:

1. There are no transport processes except the passive processes of diffusion and advection (in the assimilation and transpiration stream). This means that any active transport, e.g., with carriers, is excluded.
2. The partitioning between plant tissue and aqueous solution is driven by the lipid and the water content of the plant and the lipophilicity of the chemical (expressed as K_{OW}); this excludes processes like the "ion trap" due to dissociation of chemicals.

Some reflections and experimental studies indicate the correctness of these assumptions (for anthropogenic nondissociating organic chemicals only), but there is no claim of accuracy for all xenobiotics.

Specific carrier, transport, or exclusion systems are only likely for those substances that have been present in the environment of the plants for a long time or for those compounds that are "confounded" with ones that had an effect during evolution. As the name indicates, xenobiotic chemicals usually do not belong with these compounds. Their uptake generally should be of a passive

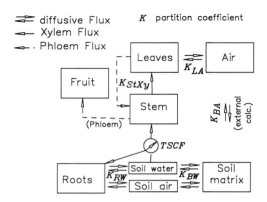

FIGURE 1.   Organization of the model PlantX. See Table 1 for definitions. (Reprinted with permission from *Environmental Toxicology and Chemistry,* Volume 13(13), Stefan Trapp, Craig Mc Farlane, Michael Matthies, "Model for Uptake of Xenobiotics into Plants: Validation with Bromacil Experiments." Copyright 1994, SETAC.)

type, i.e., via diffusion or cotransport with the plant saps. This consideration is the basis for the first assumption.

B. PARTITION COEFFICIENTS FOR PLANT TISSUE

Partition coefficients between plant tissue and other environmental media play a key role for the description of the behavior of organic chemicals in plants.

In roots exposed to contaminated aqueous solution the concentration increases at first very quickly, but then reaches a value that is rather constant over a longer time period. This value was described as the root concentration factor (RCF, units = mg chemical per g fresh plant to mg chemical per ml external solution).[16-18]

Briggs et al.[18] found a correlation between the RCF of barley roots and the lipophilicity of chemicals. The authors assumed that the RCF describes partitioning between the external solution and an aqueous phase plus a hydrophobic (lipid) phase in the roots. Following this argumentation (which is the equivalent of the second assumption above), the basic equation for sorption to plant tissue may be formulated (here for roots) as

$$K_{RW} = (W_R + l_R \cdot K_{OW}^b) \cdot \rho_R/\rho_W \qquad (3.1)$$

where K_{RW} is the partition coefficient of roots to external solution (units = mass per volume to mass per volume), W_R is the water content of the roots (mass/mass wet weight), l_R is the lipid content (mass/mass wet weight), b is a correction exponent for differences between plant lipids and octanol, ρ_R is the density (mass/volume) of the fresh root, and ρ_W is the density of the external solution rsp. water.

Luc Pussemier measured the uptake of 12 ^{14}C-labeled substituted
N-methyl-phenyl carbamates into bean plants.[19] The results were as follows:

- The correlation between adsorption and K_{OW} was 0.9 (significant on the 0.1%
  Niveau).
- The adsorption isotherm was linear and followed the Freundlich equation
  (Equation 2.6).

The analysis of the sorption experiments was done with Equation 3.1.
Water content of roots and stem was approximately 85%; lipid content (measured) was 1.1%. The fit of the correction exponent b gave

$$K_{RW} = 0.85 + 0.011 \cdot K_{OW}^{0.75} \tag{3.2}$$

The equation of Briggs et al.,[18] formulated in that manner, is

$$K_{RW} = RCF \cdot \rho_R/\rho_W = (0.82 + 0.03K_{OW}^{0.77}) \cdot \rho_R/\rho_W \tag{3.3}$$

If ρ_R and ρ_W are equal, then K_{RW} is identical to the RCF. The value 0.82
represents the water content of the root and 0.03 the lipid content. These
parameters were not determined explicitly, so this formulation cannot be
verified with the data of Briggs et al. However, the value of b (0.77) is very
similar to the value above (0.75) in Equation 3.2. Most likely, the difference
between Equations 3.2 and 3.3 is caused by differences in the lipid content
(1.1% and 3.0%, respectively) between barley roots and bean roots.

The advantage of the formulation of partitioning to plant tissue with
Equation 3.1 is the transferability between plant species. Water and lipid
contents are parameters that can be determined easily. Differences between
plant lipids (which surely exist) may be adjusted with the exponent b. Values
for b were found to be 0.77 (barley roots[18]), 0.75 (bean roots and stems[19]), 0.95
(barley shoots[20]) and 0.97 (citrus cuticles[21]).

Partition coefficients between plant tissue and water are needed by the model
for roots to soil solution (K_{RW}), stem to xylem sap (K_{StXy}), and leaves and cuticles
to water (K_{LW} and K_{CW}, respectively). They are all calculated with Equation 3.1.

C. UPTAKE AND TRANSPORT KINETICS

Beneath the partitioning that describes the concentration ratio in equilibrium,
transport kinetics may be of high importance. Following the first basic assumption of Section III.A, only passive transport mechanisms shall be considered.

1. Diffusive Uptake into Roots

As pointed out, chemical uptake from the external (soil) solution into the
roots is driven by two forces: diffusion and mass flow. Diffusion can occur as

well in air as in water-filled soil pores. The root has a more or less cylindrical shape. A solution of Fick's first law of diffusion for a cylindrical surface is given in Reference 22:

$$g_R \cdot A = 2 \cdot L \cdot \pi \cdot D_{eff}/[\ln(R_2/R_1)] \tag{3.4}$$

where g_R is the conductance between soil and root (m/s), A is the root surface (m²), L is the root length (m), D_{eff} is the effective diffusion coefficient (m²/s), and R_1 is the root radius. $R_2 - R_1$ is the diffusion length between root and soil matrix. Hereby R_2 corresponds to the radius of a deficiency zone around the root. Clearly, R_2 should be very small when the root is growing into a new soil segment. Then the uptake is fast. However, at the same time, the concentration in the rhizosphere decreases. This leads to a zone of decreased chemical concentration around the root. For phosphorus a deficiency zone of 1.5 mm in agricultural soils was found; for potassium this zone was 3 mm.[23] Both are nutrients and are actively taken up by the plants. Hence, a default value for the diffusion pathway of around 1 mm ($R_2 - R_1$) seems to be justified.

The effective diffusion coefficient, D_{eff}, is the molecular diffusion coefficient D lowered by a tortuosity or labyrinth factor that takes into account the microgeometry of the soil structure. The tortuosity factor can be estimated from the pore volume (Millington and Quirk, cited in Reference 12):

$$D_{Weff} = (\theta^{10/3})/(\varepsilon^2) \cdot D_W \tag{3.5}$$

$$D_{Aeff} = [(\varepsilon - \theta)^{10/3}]/(\varepsilon^2) \cdot D_G \tag{3.6}$$

D_{Weff} and D_{Aeff} are the effective diffusion coefficients in water-filled and air-filled pores, respectively, θ is the fraction of water-filled pores, $\varepsilon - \theta$ is the fraction of air-filled pores, ε is the total porosity, and D_W and D_A are the molecular diffusion coefficients in water and air, respectively. The chemical flux is calculated by multiplying Equation 3.4 by the concentration gradient between soil and root. Soil, root, and pore water are different media. Because only that part of the chemical present in the pores can diffuse, the partition constants must be considered. C_W is the concentration of the chemical in the soil water and may be calculated from $C_W = C_B/K_{BW}$. Hereby C_B is the bulk soil concentration (kg/m²) and K_{BW} is the soil/water partition coefficient. For diffusion in water-filled pores it follows that

$$N_{WR} = g_W \cdot A \cdot (C_W - C_R/K_{RW}) \tag{3.6}$$

where

$$g_W \cdot A = 2 \cdot L \cdot \pi \cdot D_{Weff}/[\ln(R_2/R_1)] \tag{3.7}$$

N_{WR} is the net flux of the substance by diffusion between soil and root in water-filled pores, C_R is the concentration in roots (kg/m^3) and K_{RW} is the partition coefficient between roots and water.

The diffusive net flux in air-filled pores, N_{AR}, is analogous:

$$N_{AR} = g_A \cdot A \cdot K_{AW} \cdot (C_W - C_R/K_{RW}) \qquad (3.8)$$

where

$$g_A \cdot A = 2 \cdot L \cdot \pi \cdot D_{Aeff}/[\ln(R_2/R_1)] \qquad (3.9)$$

The relative importance is mainly determined by the air/water partition coefficient, K_{AW} (the dimensionless Henry's law constant). Because diffusion in gases is ca. 10^4 times faster than in liquids, diffusion in air pores becomes relevant only for ca. $K_{AW} > 10^{-5}$.

The total diffusive net flux of a substance between soil and roots by diffusion, N_{DR}, is the sum of N_{AR} and N_{WR}:

$$N_{DR} = N_{AR} + N_{WR} =$$
$$2 \cdot L \cdot \pi/[\ln(R_2/R_1)] \cdot (K_{AW} \cdot D_{Aeff} + D_{Weff}) \cdot (C_W - C_R/K_{RW}) \qquad (3.10)$$

This estimation method for uptake and partitioning is probably realistic only for the root cortex of intact roots, not for the central cylinder that is separated by the endodermis from the external solution. The inner root might — similar to the stem — exchange at least partially with the transpiration stream. No distinction will be made here between inner and outer root (rsp. cortex and central cylinder) because the volume fractions (and some other data) are usually not given *in situ*. Equation 3.10 gives an upper limit for diffusive uptake into bulk root and is used for modeling.

2. Uptake and Transport with the Transpiration Stream

Mass flow of solute with the transpiration stream, N_T (kg/s), depends on the flow of transpired water, Q_W (m^3/s):

$$N_T = Q_W \cdot C_W \qquad (3.11)$$

To enter the xylem, the chemical must pass the symplast of the endodermis. Chemical entry has been shown to correlate with its lipophilicity. The concentration ratio between transpiration stream in the xylem and external solution is expressed as the TSCF (transpiration stream concentration factor). Briggs et al.[18] found that for barley plants

$$TSCF = 0.784 \cdot \exp[-(\log K_{OW} - 1.78)^2/2.44] \qquad (3.12)$$

The equation describes an optimum curve with a maximum TSCF of 0.784 at log K_{OW} = 1.78. It also can be considered that TSCF = 1 − σ, where σ is a reflection coefficient. The concentration of a chemical in the xylem solution, C_{Xy} (kg/m³), is then

$$C_{Xy} = TSCF \cdot C_W \qquad (3.13)$$

and the translocation with the transpiration stream within the xylem into the stem, N_{TSt} (kg/s), is

$$N_{TSt} = Q_W \cdot C_{Xy} = Q_W \cdot TSCF \cdot C_W \qquad (3.14)$$

The fraction of the chemical that enters the plant with the transpiration stream but is reflected at the endodermis remains in the roots:

$$N_{TR} = N_T - N_{TSt} = Q_W \cdot (1 - TSCF) \cdot C_W = Q_W \cdot \sigma \cdot C_W \qquad (3.15)$$

where N_{TR} is the mass flow of chemical (kg/s) with the transpiration water remaining in the roots.

For the mass balance of the portion of chemical being translocated within the xylem into stem and leaves, the following simplifying assumptions are made:

1. When entering the stem, the transpiration stream concentration, C_{Xy} (kg/m³), is determined by the concentration in soil water, C_W, and the TSCF (Equation 3.13).
2. When leaving the stem, the concentration in the transpiration stream, C'_{Xy} (kg/m³), is in equilibrium with the concentration in the stem, C_{St} (kg/m³):

$$C'_{Xy} = C_{St}/K_{StXy} \qquad (3.16)$$

K_{StXy} is the partition coefficient of stem to xylem (rsp. water). The flux in the xylem with the transpiration stream into the stem, N_{TSt}, is (Equation 3.14):

$$N_{TSt} = Q_W \cdot TSCF \cdot C_W \qquad (3.17)$$

and the flux in the xylem out of the stem rsp. the flux into the leaves with the transpiration stream, N_{TL}, is

$$N_{TL} = Q_W \cdot C_{St}/K_{StXy} = Q_W \cdot C'_{Xy} \qquad (3.18)$$

and is a loss for the stem.

3. Transport with the Assimilation Stream

Bromilow and Chamberlain[24] stated that "even compounds that move only in the apoplast can nonetheless usually enter the symplast quite freely, and all phloem-mobile compounds can also move in the xylem". Of course, this holds only for anthropogenic organic chemicals that are not, like sugars, actively loaded into the phloem. Because the mass flow in the xylem is at least one order of magnitude higher, a significant transport of chemical from leaves to roots in the phloem rarely occurs when the flow directions of xylem and phloem are opposite. Due to this, passive phloem transport of most nondissociating organic chemicals into roots should be insignificant, while transport into the fruit could be more important and is the only one considered here.

The phloem flux, Q_P (assimilation stream), into the fruit can be approximately calculated by the following assumption: in fruits, all dry mass arrives via the phloem. Phloem sap has a high dry mass concentration, about 10%.[25] Therefore, if all translocated dry mass stays in the fruit, the phloem flux Q_p (m^3/s) is

$$Q_P = (1 + a) \cdot (1 - W_F)/(1 - W_P) \cdot dV_F/dt \qquad (3.19)$$

where W_F and W_P are the water fractions of fruit and phloem sap, respectively, and dV_F/dt is the change in the fruit volume rsp. the growth in the period dt (m^3/s). The fruit uses a part of the dry mass (mainly sugars) for respiration.[26] This is expressed with the factor "a". For practical reasons the value of "a" is set to zero.

Although phloem loading with sugars occurs in the leaves, phloem concentrations of nonionized chemicals change toward the equilibrium with the stem.[27] Thus, for the calculation of the phloem flux, assumptions are made which are similar to those for the xylem flux:

1. In the leaf, the phloem sap is in equilibrium with the leaves, and the flux is from leaves into the stem:

$$N_{PSt} = Q_P \cdot C_L/K_{LW} \qquad (3.20)$$

   where N_{PSt} is the flux of the chemical within the phloem from the leaves to the stem (kg/s), C_L is the concentration in the leaves (kg/m^3), and K_{LW} is the partition coefficient between leaves and water rsp. assimilation stream (calculated from Equation 3.1).

2. In the stem, the concentration of the chemical in the phloem sap is equilibrated with the stem and comes to the same concentration as within the xylem sap (when leaving the stem). The flux of chemical from stem to fruit, N_{PF}, within the phloem (kg/s) is then

$$N_{PF} = Q_P \cdot C_{St}/K_{StXy} = Q_P \cdot C'_{Xy} \qquad (3.21)$$

4. Exchange between Leaves and Atmosphere

From leaves, substances can volatilize. When this process is slower than the vaporization of water, an accumulation in the leaves can occur.

Resistance against volatilization, r (s/m), is composed of parallel resistances through stomata (r_s) and cuticles (r_c) and the series resistance through the atmosphere r_A (mainly determined by the air boundary layer resistance).[28]

a. Stomatal Resistance

The cuticle of the leaf is nearly impermeable for water, and thus from the transpiration stream the total resistance of the stomatal pathway, r_S $(= r_s + r_A)$, can be calculated:[10]

$$E = (P_L - h \cdot P_A)/r_S \tag{3.22}$$

where E is the amount of water vapor leaving a leaf per unit area and per unit time $(kg \cdot s^{-1} \cdot m^{-2})$ and is calculated from the total transpiration stream Q_W and the leaf area A, P_L is the saturation concentration of water vapor in the leaf at leaf temperature (kg/m^3); assumed to be approximately equal to P_A, P_A is the saturation concentration of water vapor in the atmosphere at air temperature, h is the relative humidity (%/100), and r_S is the total stomatal resistance for water vapor (s/m). P_A for the corresponding temperature may be looked up in tables or calculated from the empirical Magnus equation,[29] as is done automatically in the program PlantX.

From the stomatal resistance to water vapor, r_{Sw}, the resistance for any substance, r_{Sx}, can be approximately estimated by use of the molecular weight MW (g/mol):

$$r_{Sx}/r_{Sw} = (MW_x/MW_w)^{0.5} \tag{3.23}$$

where w is the index for water and x is the index for the substance. The conductance (m/s) of the stomatal pathway, g_S, is the inverse of the resistance r_S (s/m); $g_S = 1/r_S$.

b. Cuticular Resistance

In contrast to water, organic chemicals may well volatilize via the cuticular pathway.[30] The diffusion takes place in the cuticle, with air on one side and leaf interior on the other. The concentration gradient $\Delta C/\Delta x$ (Δx is the effective diffusion length in meters) has to be defined taking this into consideration. When the concentration on the outer side of the cuticle is in equilibrium with the air, then

$$C_C = C_A \cdot K_{CA} = C_A \cdot K_{CW}/K_{AW}$$

where C_C is the concentration in the cuticle, C_A is the air concentration, K_{CA} is the partition coefficient between cuticle and air and is equal to K_{CW}/K_{AW}, K_{CW}

is the partition coefficient between cuticle and water, and K_{AW} is the partition coefficient between air and water. Next, the concentration of the leaf interior is assumed to be in equilibrium with the cuticle:

$$C_C = C_L \cdot K_{CW}/K_{LW}$$

The concentration difference that is the driving force for diffusion between inner and outer cuticle surfaces is now defined:

$$\Delta C_C = C_A \cdot K_{CW}/K_{AW} - C_L \cdot K_{CW}/K_{LW} \qquad (3.24)$$

Diffusive flux across membranes is usually expressed as a function of area A (m^2), diffusion coefficient D (m^2/s), and concentration gradient. The flux across the cuticle, N_C (kg/s), would then be

$$N_C = A \cdot D \cdot \Delta C_C/\Delta x \qquad (3.25)$$

It is advantageous to use permeances P (m/s) instead of diffusivities because the effective thickness of biological material is difficult to determine.[30] With $P = D \cdot K_{CW}/\Delta x$ (for P, concentration in water is the driving force of diffusion), it follows from Equations 3.24 and 3.25 that

$$N_C = A \cdot P \cdot (C_A/K_{AW} - C_L/K_{LW}) \qquad (3.26)$$

Kerler and Schönherr[21] showed that K_{CW} and P for isolated *Citrus* cuticles are closely related to the K_{OW}:

$$\log K_{CW} = 0.057 + 0.970 \cdot \log K_{OW} \quad n = 50, r = 0.987 \qquad (3.27)$$

$$\log P = -11.2 + 0.704 \cdot \log K_{OW} \qquad r = 0.91 \qquad (3.28)$$

These equations are used in the model. New results indicate much higher permeances rsp. conductances for isolated *Capsicum* cuticles (deduced from Reference 31):

$$\log P = -3.47 - 2.79 \log MW + 0.970 \log K_{OW} \qquad (3.29)$$

Equation 3.28 may be seen as the lower limit of cuticular permeances, Equation 3.29 as the upper limit (J. Schönherr, personal communication).

From the permeance of the cuticle, P_c, the conductance g_c (m/s) is derived:

$$g_c = P/K_{AW} \qquad (3.30)$$

For g_c, concentration in air is the driving force of diffusion. The flux of chemical across the cuticle, N_C (kg/s), finally is

$$N_C = A \cdot g_c \cdot (C_A - C_L \cdot K_{AW}/K_{LW}) \hspace{2cm} (3.31)$$

c. Air Boundary Layer Resistance

The chemical volatilizing from the leaf additionally has to overcome the air boundary layer. An estimate of the conductance between leaf surface and free atmosphere, g_A (the inverse of r_A), is $5 \cdot 10^{-3}$ m/s for chemicals with a molecular weight of 300 g/mol.[28] For other molecular weights this value may be adjusted (Equation 3.23). Whereas this value is already considered for the total resistance of the stomatal way (Equation 3.22), it must be added for the cuticular way. The total conductance, g_C (m/s), of the cuticular way is then

$$g_C = 1/(1/g_c + 1/g_A) \hspace{2cm} (3.32)$$

d. Total Flux between Leaf and Atmosphere

Together with the stomatal transfer acting in parallel, the total flux between leaf and air, N_{LA} (kg/s), is

$$N_{LA} = A \cdot (g_C + g_S) \cdot (C_A - C_L \cdot K_{AW}/K_{LW}) \hspace{2cm} (3.33)$$

The flux is diffusive and can occur in both directions, depending on the concentration gradient. The advantage of this calculation method is that it requires only a few input parameters (humidity, temperature, transpiration, leaf area, molecular weight, log K_{OW} and K_{AW}). Furthermore, see Chapter 6 in this book, which gives an excellent overview of this topic.

5. Metabolism

Plants are known to be very reactive environments for xenobiotic chemical metabolism, although not much is known about rates. In contrast to animals, plants usually have no excretion system (except the volatilization of fugitive substances and excretions of roots). After the uptake of substances, transformation reactions may occur (oxidation, reduction, and hydrolysis) followed by conjugation reactions and the building of bound residues, e.g., via glucoside complexes.[32] Conjugates are often deposited in vacuoles and cell walls. Because metabolite character cannot be predicted yet, metabolite fate is not dealt with in this model. Metabolism rates are required as input and must be determined outside the model. First-order reaction is assumed with rate constants λ (1/s). Although a value may be entered for each plant compartment, usually one value for all will be used. The model calculates the amount of metabolites formed, and in the case of ^{14}C experiments the simulated metabolite concentration may be added to the concentration of the parent compound for a comparison to the experimental results. Hereby metabolites are assumed to be immobile and to stay in the plant organ where they were formed (this holds for bound residues, but is questionable for mobile polar metabolites).

D. EQUATION SYSTEM OF THE MODEL PLANTX

The single processes combined give a model for the calculation of uptake, translocation, accumulation, metabolic conversion, and volatilization of anthropogenic organic chemicals in plants. Four compartments are considered, namely roots, stem, leaves, and fruits. They are assumed to be homogeneously mixed. In the following, the equations describing the mass balance of a chemical in the plant are given. The symbols used in the equations are explained in Table 1.

Table 1. Table of symbols

(I)	A	= area (m^2)
(D)	b	= exponent to correct differences between plant lipids and n-octanol (-)
(I,C)	C	= concentration of chemical (kg/m^3) (air, soil: input)
(C)	D	= diffusion coefficient (m^2/s)
	D_{eff}	= effective diffusion coefficient (m^2/s)
(C)	E	= amount of water vapor leaving a leaf per unit area and per unit time ($kg \cdot m^{-2} \cdot s^{-1}$)
(C)	g	= conductance (m/s) (driving force is concentration in air)
(I)	h	= relative humidity (%)
	K_{IJ}	= partition coefficient (mass/volume to mass/volume) between phases I and J, e.g.
(I)	K_{OW}	= partition coefficient between n-octanol and water
(I)	K_{AW}	= partition coefficient between air and water
		= dimensionless Henry's law constant
(C)	K_{LW}	= partition coefficient between leaf and water (C_R/C_W), wet weight
(C)	K_{RW}	= partition coefficient between root and water (C_R/C_W), wet weight
(C)	K_{BW}	= partition coefficient between bulk soil and water (wet weight)
(C)	K_{StXy}	= partition coefficient between stem and xylem (C_{St}/C_{Xy})
(I)	MW	= molecular weight (g/mol)
(I)	l	= fraction of lipids (mass/mass wet weight)
(C)	L	= length (m)
(C)	N	= flux of chemical (kg/s)
(I)	OC	= organic carbon content of soil (mass per mass dry weight)
(C)	P_A	= saturation concentration of water in air (kg/m^3)
(C)	P_L	= saturation concentration of water in leaf (kg/m^3)
(C)	P	= permeance of chemical through the cuticle (m/s) (driving force is concentration in water)
(I,C)	Q	= water flow rate (m^3/s)
(C)	r	= resistance (s/m)
(I,D)	R	= radius (m)
(I)	t	= time (s)
(I)	T	= temperature (K)
(C)	TSCF	= transpiration stream concentration factor
(I)	V	= volume (m^3)
(I)	W	= water content (vol/vol)
(C)	σ	= reflection coefficient (= 1 − TSCF)
(I)	λ	= rate constant for metabolism (1/s)

Indices:

A	=	air	W	=	water (including soil solution)
R	=	root	OW	=	octanol-water
L	=	leaf	Xy	=	Xylem
B	=	bulk soil	P	=	phloem
St	=	stem	C	=	cuticle
F	=	fruit	T	=	transpiration stream

Note: (I) = input data; (D) = default value; (C) = internal calculation. Units: SI (m, kg, s).

Equation System PlantX

<u>Roots</u>: mass change =
- ±   Diffusion from soil to roots (N_{DR}, Equation 3.10)
- +   Mass flow with the transpiration stream (N_{TR}, Equation 3.15)
- −   Metabolism

$$V_R \cdot (dC_R/dt) = 2 \cdot L \cdot \pi/[\ln(R_2/R_1)] \cdot (K_{AW} \cdot D_{Aeff} + D_{Weff}) \cdot$$
$$(C_W - C_R/K_{RW}) + Q_W \cdot (1 - TSCF) \cdot C_W - \lambda_R \cdot V_R \cdot C_R$$

<u>Stem</u>: mass change =
- +   Mass flow with the transpiration stream from soil (N_{TSt}, Equation 3.17)
- −   Mass flow with the transpiration stream to leaves (N_{TL}, Equation 3.18)
- +   Mass flow with the assimilation stream from leaves (N_{PSt}, Equation 3.20)
- −   Mass flow with the assimilation stream to fruits (N_{PF}, Equation 3.21)
- −   Metabolism

$$V_{St} \cdot (dC_{St}/dt) = Q_W \cdot (C_W \cdot TSCF - C_{St}/K_{StXy}) + Q_P \cdot$$
$$(C_L/K_{LW} - C_{St}/K_{StXy}) - \lambda_{St} \cdot V_{St} \cdot C_{St}$$

<u>Leaves</u>: mass change =
- +   Mass flow with the transpiration water from stem (N_{TL}, Equation 3.18)
- ±   Diffusive flux from/to air (N_{LA}, Equation 3.33)
- −   Mass flux with the assimilation stream to the stem (N_{PSt}, Equation 3.20)
- −   Metabolism

$$V_L \cdot (dC_L/dt) = Q_W \cdot C_{St}/K_{StXy} + A \cdot (g_C + g_S) \cdot$$
$$(C_A - K_{AW} \cdot C_L/K_{LW}) - Q_P \cdot C_L/K_{LW} - \lambda_L \cdot V_L \cdot C_L$$

<u>Fruits</u>: mass change =
- +   Mass flow with the phloem from stem (N_{PF}, Equation 3.21)
- −   Metabolism

$$V_F \cdot (dC_F/dt) = Q_P \cdot C_{St}/K_{StXy} - \lambda_F \cdot V_F \cdot C_F$$

<u>Growth</u>:
The growth of plants has a diluting effect. From the law of mass conservation it follows that

$$C_0 \cdot V_0 = C_t \cdot Vt \qquad (3.34)$$

where C_0 and V_0 are concentration and volume at time 0, respectively, and C_t and V_t are concentration and volume at time t, respectively.

E. NUMERICAL SOLUTION METHOD

The system of differential equations is solved with a one-step Euler solution scheme. A trivial correction method for numerical errors is used: as soon as negative concentrations occur (this is usually the case when the numerical integration fails), the time step is divided by the factor 2 and the integration is repeated. The model is programmed in FORTRAN77 (program PlantX); a program in C is available too, as part of the program package E4CHEM.[33] The model can be applied to plants growing in soil or solution, and the connection to a soil model is shown in Chapter 8 of this book.

An advantage of this model is its numerical stability (due to the separation of root and stem mass balances) and the need for only a few common input parameters (Table 1). All chemical parameters (partition coefficients, exchange rates) are calculated internally from the minimum data K_{OW}, dimensionless Henry's law constant K_{AW}, and molecular weight MW. This allows simulations for a large number of chemicals. Only the metabolism rate is a plant- and chemical-specific input parameter that must be either fitted or measured since no estimation method is available.

F. ANALYTICAL SOLUTION METHOD

The mass balance equation for uptake into shoots is only one single differential equation. Under constant conditions, this equation is easily solved analytically. Results from that equation can be calculated with a table calculator, too, and thus might be of practical importance.

The mass balance for the shoots (aerial plant parts) is:

Change of chemicals mass in aerial plant parts =
+   Flux from soil to shoots
±   Flux into/from the air
−   Metabolism

or, expressed in mathematical terms and for dC_L/dt:

$$dC_L/dt = Q_W \cdot TSCF \cdot C_W/V_L + A \cdot (g_C + g_S) \cdot (C_A - K_{AW} \cdot C_L/K_{LW})/V_L - \lambda \cdot C_L$$

Taking concentrations in soil solution and air (C_w and C_A, respectively) as constants yields a linear differential equation of the general form $dy/dt + ay = b$, where

$$y = C_L$$
$$a = A \cdot (g_C + g_S) \cdot K_{AW}/(K_{LW} \cdot V_L) + \lambda$$
$$b = Q_W \cdot TSCF \cdot C_W/V_L + A \cdot (g_C + g_S) \cdot C_A/V_L$$

Without growth and for a given $C_L(0)$ the analytical solution for $C_L(t)$ is

$$C_L(t) = C_L(0) \cdot e^{-at} + b/a \cdot (1 - e^{-at}) \qquad (3.35)$$

The steady-state concentration $C_L(\infty) = b/a$. The analysis of the coefficient "a" gives the relative importance of elimination pathways, whereas "b" describes the uptake pathways. The time required for the system to reach (95% of) the steady-state is calculated from $C_L(t)/C_L(\infty)$ ca. 0.95. It follows that $1 - e^{-at} = 0.95$ and $t = -\ln 0.05/a$.

G. LIMITATIONS

Due to its simplicity the model also has limitations. For example, the model was primarily designed for uptake from soil or solution. Uptake from air into leaves is possible, but there is no mechanism provided for translocation into the roots. The behavior of dissociating chemicals can be modeled, but the values for K_{RW}, TSCF, K_{StXy}, K_{LW}, and phloem concentration have to be defined in a different manner (see Chapter 3). It must be kept in mind that plants are strongly idealized by the model approach. The compartments (roots, stem, leaves, and fruits) are considered to be homogeneously mixed. In the case of larger plants, however, in reality there may be considerable differences between the basal and upper parts of these compartments. Xylem transport into fruits and the exchange between fruits and air is neglected.

Usually, the computations for the behavior of a chemical for one vegetation period require only a few seconds (IBM-PC 386). In the case of high exchange rates (e.g., for highly fugitive substances), the numerical solution is difficult and time steps must be very short. Then computations can take up to 1 h or more.

IV. COMPARISON OF MODEL PREDICTIONS WITH EXPERIMENTAL STUDIES

The purposes of this chapter are to test by a few examples whether the model is able to predict the behavior of chemicals correctly and to prove the reliability of the model. The experiments were carried out by Pussemier and Mc Farlane, who kindly supplied their results for this study. They are shown in detail: maybe they can serve one day for a validation of models developed in the future.

A. UPTAKE OF CARBOFURAN INTO BEAN PLANTS

Methods and materials are documented in Reference 19. The plant species used was *Phaseolus vulgaris* var. *Prozessor* (beans). The plants had been growing in hydroculture for 8 days and had the first two primary leaves. For up to 72 h the roots were exposed to 50 ml of a solution containing 100 µg/l Carbofuran (2,3-dihydro-2,2-methylbenzofuranyl-*N*-methyl-phenylcarbamat). Carbofuran is a systemic insecticide with a log K_{OW} of 1.82 (measured by high-pressure liquid chromatography [HPLC]).

At six different time periods plants were harvested and rinsed. The amount of water transpired was determined. Plants were divided into roots, stem, and leaves and then oxidized. The $^{14}CO_2$ was measured in a liquid scintillation counter. The artificial day/night rhythm was 16 h to 8 h. Growth of the plants was negligible. The concentration of Carbofuran in roots and stems of the bean plants quickly achieved a certain value which remained more or less unchanged afterward. In the leaves, the radioactivity increased proportionally to the amount of water transpired (Table 2).

The model is used to simulate the transient behavior of Carbofuran. The TSCF of Carbofuran calculated with Briggs' equation is 0.783. The actual TSCF of Carbofuran in the bean plants used can be determined by dividing the amount in the stem and leaves by the transpired volume of water times the solution concentration. That gives a TSCF (Carbofuran, bean plant) of 0.66. No further fit was done to carry out the simulation. Diffusive uptake into roots was neglected in this particular simulation. As can be seen in Figure 2, there is a good agreement between the dynamic simulation and the experimental results.

The behavior of Carbofuran can be interpreted by use of the model: at first, the chemical is taken up simultaneously into the roots and the xylem. From the xylem, the chemical partitions into the stem. Soon after, the stem concentration gets closer to equilibrium with the xylem sap, and the chemical is translocated to the leaves. From this time forward, the concentration in the leaves increases proportionally with the amount of water transpired. Carbofuran rarely volatilizes ($K_{AW} = 3.1 \cdot 10^{-7}$)[34] and no metabolism has been assumed.

B. UPTAKE OF BROMACIL UNDER VARYING ENVIRONMENTAL CONDITIONS

Although the preceding simulation showed very clearly the importance of the translocation with the transpiration stream and (in the absence of elimination) the accumulation in leaves, it lacked evaluation of environmental variations and various plant parameters. Experiments under controlled conditions have been carried out by Mc Farlane and his co-workers.

Table 2.   Uptake of Carbofuran with time and transpired water

Time (h)	Transpiration (ml)	Amount incorporated (mg)			Weight (g)		
		Stem	Root	Leaves	Stem	Root	Leaves
5	1.05	21	31	71	0.32	0.78	2.48
8	1.55	25	75	78	0.28	0.82	2.60
24	3.35	29	63	163	0.28	0.67	2.35
30	4.15	23	65	257	0.30	0.75	2.32
48	7.1	20	67	415	0.42	0.94	2.80
72	9.7	32	91	607	0.35	0.70	2,71
Mean	0.135 ml/h	25	65.3		0.325	0.777	2.54

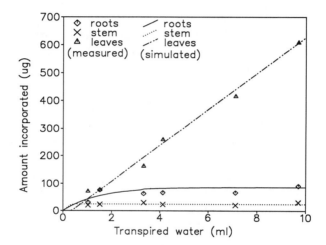

FIGURE 2.   Transient simulation of Carbofuran.

1. Experimental Methods

Some aspects of the plant uptake studies were common to all experiments. Plants were grown in a hydroponic nursery[35] in a glasshouse in a modified, half-strength Hoagland's nutrient solution with a pH of 6.0 and an electrical conductivity of 1.2 dS/m. Plants of similar size were transferred to specially built plant exposure chambers[36] and were acclimated (typically for 3 days) to the conditions of the controlled environment (photosynthetic photon flux [PPF] = 350 μmol $\cdot$ s^{-1} $\cdot$ m^{-2}, air temperature = 23°C, air turbulence = 0.6 m $\cdot$ s^{-1}, CO_2 = 15.65 mmol/m^3) prior to treatment with the test chemical. The total solution volume in each chamber was 6.5 l; this was circulated at 11.4 l/min. ^{14}C ring-labeled chemicals were added to the hydroponic solution through an injection port, which guaranteed rapid mixing and, thus, uniform treatment of each plant in the study. The shoot environment was enclosed within a clear Teflon® tank in which the air was stirred and the CO_2 and H_2O vapor were controlled. Plant transpiration and photosynthesis were measured and the data were used to verify the health and normal functioning of the plants.

Bromacil uptake studies were conducted in two different formats to examine the versatility of the system and the responsiveness of the model. The first included a single addition of Bromacil to the nutrient solution (acute exposure); the second was designed to maintain a constant exposure for the duration of the study (chronic exposure). Plants were periodically removed from each chamber, whereupon weights, areas, and ^{14}C content were determined. Attention was given to the possibility of chemical loss in drying and of incomplete combustion in the oxidation. Quality assurance tests confirmed that there was less than 1% loss in either procedure. Thin layer chromatographs were made from plant extracts and from the hydroponic solution, and nothing but parent chemical (Bromacil) could be detected, but measurable amounts of ^{14}C remained in the insoluble pellet.

2. Plant Input Data

The experiment with chronic exposure was named BROM5, the acute BROM3.[37] For both experiments, there have been three runs with a high, a moderate, and a low transpiration rate (low, moderate, and high humidity, respectively) and eight soybean plants (*Glycine max* [L.] Merr. cv. Fiskby v). The data sets used for the simulations are the averaged measured values for the eight plants (Table 3). During the low transpiration rate (low TS) BROM5 experiment the plants were grown without light.

Some input parameters were not determined, since the experiments were originally not used for the PlantX model. Estimates of these values are used. Most of them are not critical and do not influence the results. They were not varied and are the same for all plants: root radius (R_1) is 1 mm (R_2) is the default value, R_1 + 1 mm), root lipid content (l_R) is 1%, stem lipid content (l_{St}) is 3%, and leaf lipid content (l_L) is 3%. The water contents were determined for one set of plants and used for all experiments — namely, root water content (W_R) is 94.2%, stem water content (W_{St}) is 75.6%, leaf water content (W_L) is 72.7%, and fruit water content (W_F) is 77.0%. The correction exponents b needed by Equation 3.1 are 0.77 for roots and stem[18] and 0.95 for leaves.[20]

3. Chemical Input Data

Bromacil ($C_6H_{13}BrN_2O_2$) is a herbicide, mainly adsorbed through roots, used for nonselective inhibition of photosythesis and recommended for general weed control.[38] In soil, Bromacil is relatively persistent (65% metabolized within 1 year[39]). The K_{OW} is 105 (log K_{OW} is 2.02), the dimensionless Henry's law constant, K_{AW}, is $3.65 \cdot 10^{-9}$, and the molecular weight is 261.12 g/mol.

Table 3. Plant input data (averaged values)

	BROM5			BROM3[a]		
	HT	MT	LT	HT	MT	LT
Leaf area (cm²)	825	725	765	1794	1761	1578
Transpiration (ml/h)	7.3	5.7	2.0	9.5	7.6	4.4
Root mass (g)	31.5	31.1	24.7	47.9	56.5	45.5
Stem mass (g)	6.6	6.5	4.5			
Leaf mass (g)	19.9	17.3	16.7			
Initial fruit mass (g)	5.4	7.0	2.6			
Final fruit mass (g)	21.8	26.3	4.6			
Shoot mass (g)				77.0	81.4	71.2
Relative humidity (%)	50	65	85	50	55	65[b]
Solution concentration (DPM/ml)	3,250	3,250	3,250	*	*	*

Note: HT, MT, LT = experimental with high, medium, and low transpiration, respectively.
   DPM is ¹⁴C activity rsp. disintegrations per minuite.

[a] Shown later together with simulations in Figures 6 and 7.

[b] BROM3: initial values.

Source: Reprinted with permission from Environmental Toxicology and Chemistry, Volume 13(13), Stefan Trapp, Craig Mc Farlane, Michael Matthies, "Model for Uptake of Xenobiotics into Plants: Validation with Bromacil Experiments." Copyright 1994, SETAC.

4. Metabolism Rate

The metabolism rate of Bromacil in soybean during this experiment has to be determined, as pointed out above. Principally, two different methods may be applied:

- The model results can be fitted by running the model and optimizing the metabolism rate by comparison of model output and measured results.
- Since a validation allows no fit, a second method is used here to determine the metabolism rate from the measured data. If the measured concentrations are plotted vs. time, a linear increase is found, and a (nearly) straight line results that does not go through the origin (Figure 3). A regression against time is used to extrapolate the y-axis interception, where

$$Y = a + b \cdot \Delta t \tag{4.1}$$

Y is the estimate of the concentration, "a" is the y-axis intercept, "b" is the slope of the curve, and Δt is the time period. Now, if the transfer between solution and roots and stems, rsp., is very fast, the intercept "a" can be considered as the concentration of the parent compound, C_0. The difference between initial and final ¹⁴C levels then originates from the building of (immobile) metabolites with concentration C_M. This means that Y represents $C_0 + C_M$, and the slope b is $C_0 \cdot \lambda$. If the solution concentration is constant (BROM5), the metabolism half-time, $t_{1/2}$ (first order), may then be calculated from

$$t_{1/2} = \ln 2 \cdot a/b \tag{4.2}$$

The metabolism half-lives determined are shown in Table 4. The regressions have similar slopes and intercepts for all roots, independent of the transpiration,

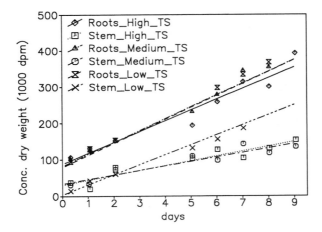

FIGURE 3. Regressions to find the metabolism rates in soybean plants. TS = transpiration; dpm = disintegrations per minute. (Reprinted with permission from *Environmental Toxicology and Chemistry,* Volume 13(13), Stefan Trapp, Craig Mc Farlane, Michael Matthies, "Model for Uptake of Xenobiotics into Plants: Validation with Bromacil Experiments." Copyright 1994, SETAC.)

and also for two of the stems. Only the low TS stem regression is unique in slope and intercept. The average half-life (without BROM5 stem low TS) is 1.8 days and serves as input for the model simulations of the BROM3 experiment.

5. Results

The values in Tables 3 and 4 are used as input data for the simulations. The model gives the simulated concentration of the total ^{14}C — that is, the concentration of the parent compound plus the concentration of the metabolites.

a. BROM5 (Chronic Exposure)

Figure 4 shows the comparison of the model predictions with the measured concentrations of the BROM5 experiment with high transpiration. In contrast, Figure 5 shows the simulation of BROM5 with low transpiration. The overall pattern is the same, but the concentration in leaves is much lower due to the lower transport in the xylem. The concentration of ^{14}C in stems is higher; this may be attributed to the fast metabolism rate. Because of the uncertainty of the metabolism rate in stems (Table 4), it was fitted with the model, giving a half-life of 0.45 days.

b. BROM3 (Acute Exposure)

Within the BROM3 experiments, only the total activity in the shoots was determined (on a wet weight basis). The comparisons between experiments and

Table 4. Estimates of the metabolic half-life ($t_{1/2}$)

Experiment	Site	Intercept a	Slope b	R^2 (%)	$t_{1/2}$ (d)
BROM5 HT	Roots	87.9	29.6	93.7	2.05
BROM5 MT	Roots	82.1	32.5	96.2	1.75
BROM5 LT	Roots	78.8	32.8	87.0	1.66
BROM5 HT	Stem	30.4	13.2	86.3	1.60
BROM5 MT	Stem	33.6	12.2	86.3	1.91
BROM5 LT	Stem	4.0	27.3	98.2	0.10

Note: HT, MT, LT = high, medium, and low transpiration, respectively.

simulations for the high and low transpiration experiments are shown in Figures 6 and 7, respectively. In any case there are small deviations, but generally the agreement is good. It must be kept in mind that the input data are averaged values, and the metabolism half-time used is the average of the BROM5 experiments.

6. Discussion

The concentration pattern in each of the Bromacil experiments is similar. The pattern can be interpreted by a look at the mass balances. The example of BROM5 high TS is shown in Table 5. Metabolism (here pertaining to building of bound residues) seems to be a very important process; the calculated amount is three quarters of the total radioactivity found after 9 days.

Following the simulations, Bromacil behaves as follows: the uptake into the roots is rapid, until the equilibrium K_{RW} (1.3 wet weight based, 22.4 dry weight based) is reached. The following, almost linear increase of activity in roots is attributed to the building of bound residues. Due to a rather high TSCF (0.766), Bromacil is quickly translocated into the xylem, from where it sorbs to the stem. A linear increase of the activity in stems occurs, again caused by metabolites. The uptake into leaves is proportional to the transpiration stream. In the leaves there is no sink: the volatilization is ineffective (the K_{AW} is very low), and the metabolism leads to a loss of Bromacil, but not of ^{14}C. The behavior in the fruit is similar, but the transport with the assimilation stream is much smaller, and so the activity there is the lowest.

The amount of bound residues of Bromacil in the soybean plant was not determined directly by measurements. In another set of experiments, carried out with eight different plant species contaminated with nitrobenzene, 30 to 70% of all ^{14}C in the roots and 15 to 40% of all ^{14}C in the shoots was insoluble (bound residues) after only 3 days.[40] This shows that plants are capable of metabolizing organic chemicals rather quickly to bound residues.

C. OUTLOOK

Some more validation studies have been carried out. They have shown that the fate of nitrobenzene in this system may be calculated correctly for some different plant species. Because nitrobenzene is rather volatile, it could be shown that the

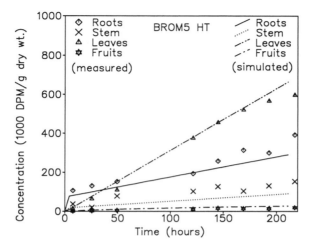

FIGURE 4. Comparison of measured and simulated concentration (^{14}C) in soybean; chronic Bromacil exposure, high transpiration (HT). (Reprinted with permission from *Environmental Toxicology and Chemistry,* Volume 13(13), Stefan Trapp, Craig Mc Farlane, Michael Matthies, "Model for Uptake of Xenobiotics into Plants: Validation with Bromacil Experiments." Copyright 1994, SETAC.)

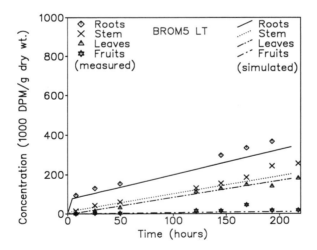

FIGURE 5. Comparison of measured and simulated concentration (^{14}C) in soybean; chronic Bromacil exposure, low transpiration (LT). (Reprinted with permission from *Environmental Toxicology and Chemistry,* Volume 13(13), Stefan Trapp, Craig Mc Farlane, Michael Matthies, "Model for Uptake of Xenobiotics into Plants: Validation with Bromacil Experiments." Copyright 1994, SETAC.)

model predicts leaf/air interaction successfully.[41] The uptake of chloroorganic compounds in another small laboratory microcosm has been investigated.[42]

All these studies were comparisons to laboratory experiments. An application to agricultural field situations is presented in Chapter 8 of this book.

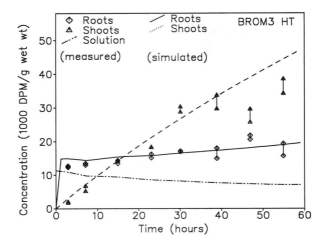

FIGURE 6. Comparison of measured and simulated concentration (^{14}C) in soybean; acute Bromacil exposure, high transpiration (HT). Measurements: maximum and minimum of three replicates. (Reprinted with permission from *Environmental Toxicology and Chemistry,* Volume 13(13), Stefan Trapp, Craig Mc Farlane, Michael Matthies, "Model for Uptake of Xenobiotics into Plants: Validation with Bromacil Experiments." Copyright 1994, SETAC.)

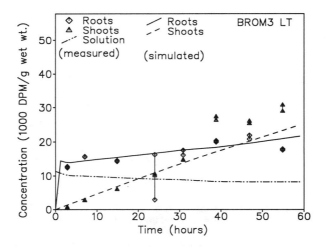

FIGURE 7. Comparison of measured and simulated concentration (^{14}C) in soybean; acute Bromacil exposure, low transpiration (LT). Measurements: maximum and minimum of three replicates. (Reprinted with permission from *Environmental Toxicology and Chemistry,* Volume 13(13), Stefan Trapp, Craig Mc Farlane, Michael Matthies, "Model for Uptake of Xenobiotics into Plants: Validation with Bromacil Experiments." Copyright 1994, SETAC.)

Table 5. Mass balance of ^{14}C BROM5 high
transpiration rate experiment after 9-day
simulation

Mass balance	1000 DPM	% of Uptake
Total uptake	4552.2	100.0
Bromacil in roots	133.9	2.9
Metabolites in roots	411.0	9.0
Bromacil in stems	32.5	0.7
Metabolites in stems	123.6	2.7
Bromacil in leaves	925.1	20.4
Metabolites in leaves	2775.0	61.1
Bromacil in fruits	25.4	0.6
Metabolites in fruits	114.1	2.5
Volatilized	11.6	0.3

Note: DMP = disintegrations per minute.

Source: Reprinted with permission from *Environmental Toxicology and Chemistry,* Volume 13(13), Stefan Trapp, Craig Mc Farlane, Michael Matthies, "Model for Uptake of Xenobiotics into Plants: Validation with Bromacil Experiments." Copyright 1994, SETAC.

V. SENSITIVITY ANALYSIS

A sensitivity analysis was conducted and yielded some insight into the general behavior of organic chemicals in the soil-plant-air system. For the sensitivity study, the same measured plant properties are used as for the model validation (BROM5). Uptake of various chemicals into roots, stem, and leaves is discussed. Hereby, the reaction of the model to a variation of input data is investigated. Experiments are cited to support the model results.

A. ROOTS

Since roots are (mathematically!) not connected with the other compartments, they can be looked at separately.

1. Will the Root be in Equilibrium?

The first interesting point is the comparison of uptake via diffusion, N_{DR} (Equation 3.10), and uptake with water, N_{TR} (Equation 3.15). In principal, a buildup of chemical at the endodermis above the equilibrium condition could occur when $N_{TR} > N_{DR}$, since the flux N_{TR} is unidirectional.

When comparing the calculated values of the diffusive flux with the mass flux (root properties— see experiment BROM5 high TS), the uptake into roots is about 95% diffusive. Only under unrealistic circumstances (root radius R_1 = 5 mm, deficiency zone radius R_2 = 10 mm; compare Section III.C.1) is the flux N_{TR} larger than N_{DR}. However, corn usually has root diameters between 0.2 and 0.3 mm.[43] It follows that an enrichment of compounds in roots above the equilibrium is very unlikely. This also holds for uptake from soil.

In experiments the contaminant concentration of roots (exposed to aqueous solution) rapidly increases to a plateau for most chemicals.[17] However, with some chemicals ^{14}C continues to increase in roots as the contaminant is metabolized to form an insoluble bond with the cell structural material. These findings support our conclusion that the diffusive flux into the roots, N_{DR}, is faster than the mass flux with the transpiration stream, N_{TR}, and the chemical can diffuse back from the root into the solution when the gradient ($C_W - C_R/K_{RW}$) becomes negative.

2. What is the Equilibrium between Root and Soil?

Another interesting aspect is the value of the root-to-soil equilibrium coefficient, K_{RB}. This is the ratio of the partition coefficient between root and water, K_{RW}, and that between bulk soil and water, K_{BW}. K_{BW} can be estimated with the methods presented in Section II of this chapter (Equations 2.7 to 2.10). Here, Equation 2.10[8] is applied to estimate the K_{OC}. Then the power on the K_{OW} is similar for soil K_{BW} (0.72) and for root K_{RW} (0.77). Subsequently, the ratio K_{RW}/K_{BW} gives relatively constant values for the hydrophobic sorption of chemicals, or, in other words, in most cases the concentration in roots will be similar to that in soil. The sorption capacity of soil organic carbon content seems to be similar to that of root lipids. This is not surprising because the organic matter in soil originates predominately from rotten plants.

Figure 8 shows results from an experimental study in a small laboratory system with seedlings of barley and cress.[42,44] The concentrations are corrected for fractions of metabolites and represent parent compounds only. As predicted (line; parameters OC is 2%, l_R is 3%), the concentration ratio is relatively constant. For hydrophilic chemicals, the root-to-soil water content ratio has increasing importance (broken part of the line). This prediction cannot be verified with this set of experiments since only lipophilic chemicals were studied. The difference between measured uptake of cress and barley is larger than that between the different chemicals. The same laboratory system (with some improvements) was used to investigate the influence of plant species on the uptake of the hydrophobic chemical hexachlorobenzene (HCB; log K_{OW} = 5.47).[45] Root/soil concentration ratios of 0.8 (barley and maize), 0.9 (oat), 3.2 (rape), 13.7 (lettuce), and 31.6 (carrot) were found. The expected value range is ca. 0.5 to 5 (Figure 9). Lettuce and carrot show considerable deviations.

B. AERIAL PLANT PARTS

Chemicals reach aerial plant organs in two ways: with the transpiration stream and from the air. The uptake from soil solution into stem and leaves is considered elsewhere in this book (Chapter 3). The exchange between air and leaves is also intensively described (Chapter 6). However, there is always competition between the translocation with the transpiration stream into aerial plant parts and the exchange of these plant parts with air.

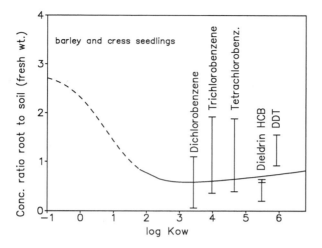

FIGURE 8.   Root-to-soil concentration ratio, (bars: measured;[44] line: predicted). HCB = hexachlorobenzene. See text for details.

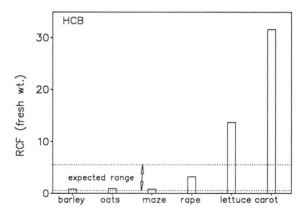

FIGURE 9.   Measured root concentration factor (RCF) for hexachlorobenzene (HCB).[45]

If the air is completely uncontaminated, the atmosphere will be a sink by which the leaf concentration is decreasing. However, in many cases pollution originates from air. Then the exchange with air can lead to contamination of plants despite the fact that the soil is clean. This might, for example, be a problem for ecologically oriented agriculture.

Often, pollution is global and long lasting. Then, under background conditions, the air/soil concentration ratio of most persistent environmental chemicals should be close to equilibrium. The relative importance of uptake from air being in equilibrium with soil is calculated using the scenario of the BROM5 experiment with moderate transpiration rates (Table 5). The percentage of uptake from air is shown in Figure 10 for a range of log K_{OW} and log K_{AW} values. As can be seen, for chemicals with a high K_{AW} (>0.001) the exchange with air is the dominating process. The same is calculated for chemicals with a high log K_{OW} (>6).

For the contamination of aerial plant parts rsp. leaves, there are two consequences:

- Chemicals being translocated well with the transpiration stream, but then volatilizing rapidly, like volatile chloroorganic chemicals, do not have a high potential for accumulation in leaves. The concentration of these chemicals in leaves depends mainly on the concentration in the air.
- Lipophilic chemicals that are not translocated in significant amounts within the plants have nonetheless a high potential for accumulation in leaves from air.

It follows that it is not enough to look at the soil when cases of plant contamination are investigated. For many classes of compounds (e.g., poly-chlorinated biphenyls [PCBs], PCA, dioxins, and volatile compounds), the air might be responsible for problems.

VI. APPLICATION EXAMPLES

In Sections IV and V the applicability of the model was shown. Now the model can be used to answer some open questions in environmental science and make predictions of pollutant behavior in the terrestrial environment. Note that the model is of an evaluative type, which means it includes assumptions and can give hypotheses but cannot prove them. Due to this, it should always be combined with experimental studies. What can be done is predicting the principal behavior and fate of chemicals and the importance of processes in the soil-plant-air system. This allows the concrete design of experiments and avoids unnecessary analytical expense.

A. BEHAVIOR OF 2,3,7,8-TETRACHLORODIBENZO-P-DIOXIN IN THE TERRESTRIAL ENVIRONMENT

1. Relevance of the Problem

2,3,7,8-Tetrachlorodibenzo-p-dioxin (2,3,7,8-TCDD) is an extremely toxic substance.[46] Furthermore, it is very persistent and lipophilic, thus accumulating in the food chain. There is considerable concern about daily intake of these (and related) compounds. In Europe, including Germany, legal standards for soil concentrations of 2,3,7,8-TCDD have been discussed.[47-49] Important

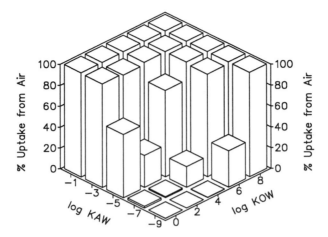

FIGURE 10. Relative importance of uptake from air; equilibrium between air and soil assumed.

for determining the acceptable level in soil is the transfer from soil to plant, because uptake into plants is the first step in the contamination of the food chain.

2,3,7,8-TCDD seems to be ubiquitously distributed all over Europe (and probably the Northern Hemisphere). Background concentrations found in soil at a rural site near Bayreuth (Germany) were around 70 pg 2,3,7,8-TCDD per kilogram soil (dry weight).[50] At a rural site in Rothamsted (southern England), recent values between 58 and 210 pg/kg were found.[50] Concentrations in plants (herbage) are of the same order of magnitude, 30 pg/kg (dry weight) in Bayreuth[50] and 32 pg/kg (dry weight) in Rothamstad.[51] Air concentrations are low; measurements of background concentrations gave 3.6 fg/m^3 (3.6 · 10^{-15} g/m^3); 75% thereof were present in the gas phase and 25% were particle bound.[52]

2. Uptake into Plants

The magnitude and pathway of uptake into plants are difficult to determine by experiments. The reason for this is not only the extreme toxicity, but also analytical difficulties due to very low concentrations and the required time scale, since exchange is slow, as well as difficulties with sorption to glass vessels, filters, etc. and the ubiquitous background contamination. Nonetheless, it has been concluded from various findings that there is evidence of uptake and translocation in plants;[53] this was not supported by a careful study with radioactive isotope-labeled 2,3,7,8-TCDD.[54]

The calculation of the mass balances of translocation in the plant and uptake from air can help in the interpretation of the environmental behavior of 2,3,7,8-TCDD. The idea here is to repeat carefully all steps necessary for the calculation of the uptake from soil and air into aerial plant parts, giving the reader an example to follow. This includes:

    a. Uptake from soil
       1. Determine concentration in soil solution
       2. Determine concentration in xylem sap
       3. Calculate mass flux into aerial plant parts
    b. Uptake from air
       1. Determine equilibrium between leaves and air
       2. Determine conductances
       3. Calculate diffusive flux

The calculation of these equations by a table calculator is a little bit troublesome, but nevertheless instructive. When all steps are considered correctly, useful estimates of the fluxes can be obtained.

a. Uptake from Soil

(1) Concentration in Soil Solution — The concentration in bulk soil, C_B, is ca. 70 pg/kg (dry weight), rsp. ca. 90 ng/m^3. The solution concentration, C_W, is C_B/K_{BW} (Equation 2.7). The K_{BW} can be estimated via $K_{BW} = K_{OC} \cdot OC \cdot \rho$ (Equations 2.8 to 2.10). An experimentally determined value of the K_{OC} is $2.5 \cdot 10^6$ cm^3/g.[55] With realistic values of OC = 2% and ρ = 1.3 g/cm^3, the K_{BW} is $6.5 \cdot 10^4$. The concentration of 2,3,7,8-TCDD in soil water C_W is then

$$C_W = C_B/K_{BW} = 90 \text{ ng/m}^3 / 6.5 \cdot 10^4 = 1.4 \text{ pg/m}^3 \qquad (6.1)$$

(2) Concentration in Xylem Sap rsp. Transpiration Stream — The concentration ratio between transpiration stream and soil solution is calculated with the TSCF (Equation 3.12). 2,3,7,8-TCDD has an extremely high log K_{OW} of around 6.76,[55] yielding a TSCF of only $3 \cdot 10^{-5}$. The concentration in the xylem sap, C_{Xy} is then (Equation 3.13):

$$C_{Xy} = \text{TSCF} \cdot C_W = 3 \cdot 10^{-5} \cdot 1.4 \text{ pg/m}_3 = 4.2 \cdot 10^{-5} \text{ pg/m}^3 \qquad (6.2)$$

(3) Mass Flux into Aerial Plant Parts — The mass flux depends on the amount of water transpired. Here an "average" plant is considered, with leaf mass of 1 kg, leaf area A of 4 m^2, transpiration Q of 0.3 l/h rsp. $8.33 \cdot 10^{-8}$ m^3/s, and leaf density of 0.5 kg/l. The mass flux with the transpiration stream into aerial plant parts, N_{TSt}, is subsequently (Equation 3.14):

$$N_{TSt} = TSCF \cdot C_W \cdot Q = C_{Xy} \cdot Q =$$
$$4.2 \cdot 10^{-5} \text{ pg/m}^3 \cdot 8.33 \cdot 10^{-8} \text{ m}^3\text{/s} = 3.5 \cdot 10^{-12} \text{ pg/s} =$$
$$3.0 \cdot 10^{-7} \text{ pg/d} = 3.0 \cdot 10^{-5} \text{ pg/100 days} \qquad (6.3)$$

Under these conditions an uptake from soil can never explain the measured (aerial) plant concentrations of ca. 30 pg/kg 2,3,7,8-TCDD (dry weight), which correspond to ca. 3 to 10 pg/kg wet weight.

The chemicals used to establish the equation for the TSCF were less lipophilic, so the result is uncertain. However, in experiments no significant translocation of 2,3,7,8-TCDD in the xylem could be detected.[54]

b. Uptake from Air into Leaves

So how does the 2,3,7,8-TCDD get into aerial plant parts? Another possible pathway is uptake from air, although background concentrations in air are very small (2.7 fg/m^3, gas phase).

(1) Equilibrium between Leaves and Air — The equilibrium between leaves and air, the K_{LA}, is derived as the ratio of the leaf/water partition coefficient, K_{LW}, and the air/water partition coefficient, K_{AW} (Equation 2.11). The leaf/water partition coefficient, K_{LW} (with 3% lipid content), is (Equation 3.1):

$$K_{LW} = 1 \cdot K_{OW}^b \cdot \rho_L/\rho_W$$
$$= 0.03 \cdot K_{OW}^{0.95} \cdot 0.5 = 4 \cdot 10^4 \qquad (6.4)$$

where ρ_L is the density of the leaf (ca. 0.5 kg/l) and ρ_W is the density of water (1 kg/l). A measured value of the K_{AW} of 2,3,7,8-TCDD is 0.0015. The K_{LA} is then

$$K_{LA} = K_{LW}/K_{AW} = 4 \cdot 10^4/0.0015 = 2.7 \cdot 10^7 \qquad (6.5)$$

This means that 2,3,7,8-TCDD has a very high potential for accumulation from air. The equilibrium concentration in leaf with background concentrations in air (2.7 fg/m^3) is

$$C_L = K_{LA} \cdot C_A = 2.7 \cdot 10^7 \cdot 2.7 \text{ fg/m}^3 = 73 \text{ ng/m}^3 = 146 \text{ pg/kg} \qquad (6.6)$$

(2) Determine Conductances — g_C and g_S are the conductances of the transfer via cuticle and stomata (m/s), respectively. Estimates of the stomatal conductance can be yielded via the resistance r (= 1/g) of the leaf against volatilization of water vapor. Values range from 40 to 1190 s/m, with an average of 650 s/m rsp. g_S of water vapor is 0.0015 m/s.[10] Adjustment for the

slower diffusivity of 2,3,7,8-TCDD is done approximately with Equation 2.15 and gives $g_{S(TCDD)}$:

$$g_{S(TCDD)} = g_{S(H_2O)} \cdot (MW_{H_2O}/MW_{TCDD})^{0.5} = 0.0015 \cdot$$

$$(18/322)^{0.5} = 0.00036 \text{ m/s} \tag{6.7}$$

The conductance of cuticles is estimated from the permeances (divided by K_{AW}) with Equations 3.30 and either Equation 3.28 (g_C for 2,3,7,8-TCDD = 0.0002 m/s) or Equation 3.29 (g_C for 2,3,7,8-TCDD = 0.08 m/s). When the air conductance is added (Equation 3.32), the g_C is 0.0002 rsp. 0.0047 m/s. The comparison to the stomatal conductance (0.00036 m/s) shows that uptake may be faster via the cuticle.

(3) Estimate of the Flux of 2,3,7,8-TCDD from Air to Leaf — The flux between leaves and air, N_{LA} is (Equation 3.33):

$$N_{LA} = A \cdot (g_C + g_S) \cdot (C_A - CL \cdot K_{LA}) \tag{6.8}$$

where A is the leaf area (4 m²). When $CL/K_{LA} \ll C_A$, at the beginning of the contamination the leaf concentration can be neglected, which simplifies Equation 6.8:

$$N_{LA} = A \cdot (g_C + g_S) \cdot C_A \tag{6.9}$$

Now the flux can be calculated easily. The result (for the lower conductance gC) is

$$N_{LA} = 4m^2 \cdot (0.00036 + 0.0002 \text{ m/s}) \cdot 2.7 \cdot 10^{-3} \text{ pg/m}^3 = 6 \cdot 10^{-6} \text{ pg/s}$$
$$= 0.5 \text{ pg/day} \tag{6.10}$$

or, with the high g_C,

$$N_{LA} = 4m^2 \cdot (0.00036 + 0.0047 \text{ m/s}) \cdot 2.7 \cdot 10^{-3} \text{ pg/m}^3 = 55 \cdot 10^{-6} \text{ pg/s}$$
$$= 4.7 \text{ pg/day} \tag{6.11}$$

The calculation for 60 days of uptake (neglecting the gradient) gives a range from 31 pg/kg leaf (fresh weight) to 283 pg/kg, which is even more than the level found in the environment. Most likely, an elimination process in the leaves exists. Indeed, an effective photodegradation has been proved experimentally with 2,3,7,8-TCDD sorbed on leaves and exposed to natural sunlight. The process is fast, with a half-life of only 1.44 days.[56]

3. Conclusions

This approximate calculation is hopefully easy to understand, it shows that the uptake of 2,3,7,8-TCDD from air is much higher than from soil and may well explain environmental concentrations.

It becomes obvious that plants are a major target and a major sink for (airborne) 2,3,7,8-TCDD and related compounds in the environment. Clearly, the concept of transfer factors between soil and plants (which is often used in hazard assessment) is obsolete when uptake is mainly from air. This should be considered in the discussion of legal standards for 2,3,7,8-TCDD and related compounds.

Nonetheless, plants grown at sites contaminated with "dioxins" could show high concentrations due to other uptake pathways. Among them are direct uptake from soil (i.e., into roots and into leaves or fruits that are in direct contact with soil) and uptake with adhered soil particles. Furthermore, direct uptake of soil — e.g., by grazing animals — is a means for the contaminants of the food chain. Therefore, a legal standard should also consider the type of agriculture.

B. UPTAKE FROM AIR VS. UPTAKE FROM SOIL

As discussed above, the air concentration might have a very significant influence on the concentration in leaves. The next example will show the difference in uptake pattern between a pure contamination of the soil and an occurrence of a pollutant in soil and in air, both in equilibrium.

1. Uptake of a Moderately Volatile Chemical from Soil

Lindane ($C_6H_6Cl_6$) is a chloroorganic insecticide with a molecular weight of 291.7 g/mol and a log K_{OW} of 4.1,[1] its K_{AW} is ca. $1.7 \cdot 10^{-5}$ (calculated from solubility and vapor pressure). Metabolism in the plant is assumed with a half-life of 100 days. The plant is similar to that used in the BROM5 high TS experiment (Section IV). In the first simulation, a bulk soil concentration of 1 µg/kg and an air concentration of 0 ng/m³ are assumed. The result is shown in Figure 11.

Uptake into roots is faster than into the leaves and leads to a steady-state concentration of ca. 50% of the soil concentration. Uptake into leaves takes much longer, lasting the whole simulation time, with a final concentration similar to that in roots.

2. Simultaneous Uptake of a Moderately Volatile Chemical from Soil and Air

In contrast to the first simulation, the air concentration is now assumed to be in equilibrium with the soil concentration (C_{Air} = 0.75 ng/m³). The result is

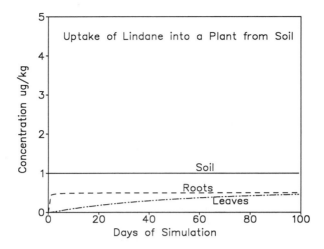

FIGURE 11. Uptake of Lindane from soil.

shown in Figure 12. Uptake into the roots remains unchanged, but uptake into leaves is nearly tenfold times higher. What happened?

In case 1 (uptake from soil), 190% of the amount found in leaves was transported to them via the transpiration stream, but 90% was volatilized into air. In case 2 (uptake from soil and air), 28% of the (much higher) amount in leaves was translocated with the transpiration stream, and 72% came into the leaves from air.

Fruits are rarely contaminated, except directly from air (not included in the model).

This simulation shows that even small concentrations in air (maybe below the detection limit) may have a significant influence on the uptake pattern of organic chemicals.

C. THE PROBLEM OF BOUND RESIDUES

While comparing the model results with the Bromacil uptake experiments, one difficulty was the building of bound residues. As pointed out earlier in this book (Chapter 4), the metabolism in plants will in many cases result in bound residues. What would happen within 100 days if the soil concentration was constant (1 µg/kg) and the plant was metabolizing Bromacil, with a half-time of 1.8 days, to bound residues? The answer is given in Figure 13.

Although the concentration of the parent compound, Bromacil, is reaching a steady-state concentration (ca. 12 µg/kg), the amount of bound residues would still increase linearly with time. The final concentration of the bound residues is more than 200 times higher than the concentration of the original

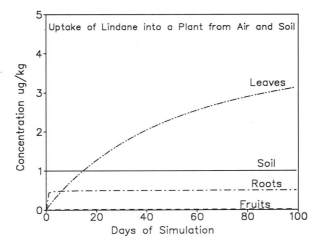

FIGURE 12.   Uptake of Lindane from air and soil.

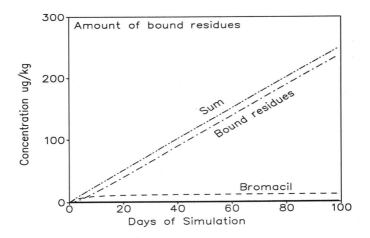

FIGURE 13.   Bound residues of Bromacil after 100 days.

substance in soil. This simulation has some unrealistic assumptions that lead to an overestimation (e.g., the concentration in soil is homogeneous and constant, and the plant is not killed by the herbicide Bromacil). It may, however, be a warning: bound residues in plants are difficult to investigate experimentally. Amount, identity, and toxicity of metabolites are mostly unknown, as are their hazards and risks.

VII. CONCLUSIONS

The manyfold processes which xenobiotic organic chemicals undergo in plants have been investigated by use of the model PlantX. The findings were compared to experimental results. Although the simulations could be carried out without model calibration, the predictions showed good agreement with the measurements. Only a few input data and a common personal computer were necessary. This allows one to carry out a large number of simulations routinely.

At present, the primary use of the model is for assisting scientists in experimental design and in finding hypotheses and for the interpretation of measured results. The next step is the simulation of field situations. Therefore, the plant exposure model needs to be combined with a soil exposure model, as described in the last chapter of this book.

Development and application of the model PlantX will hopefully lead to a better understanding of the complex processes of chemical behavior in our terrestrial environment and bring science in this area forward.

ACKNOWLEDGMENTS

Thanks to Dr. Luc Pussemier, Dr. Irene Scheunert, Dr. Rainer Schroll, and Prof. Craig Mc Farlane for letting me use their experimental results and for their valuable discussions. Thanks to Professor M. Matthies for the hint to do this work. A version of this text (in German language[41]) was reviewed by Prof. J. Schönherr, Prof. E. Sackmann, and Prof. H. Ziegler (all faculty members of the Technical University of Munich).

REFERENCES

1. Tinsley, I. (1979), *Chemical Concepts in Pollutant Behaviour*, John Wiley & Sons, New York.
2. Mackay, D. and Paterson, S. (1981), Calculating fugacity, *Environ. Sci. Technol.,* 15(9):1006–1014.
3. Lyman, W., Reehl, W., and Rosenblatt, D. (1982), *Handbook of Chemical Property Estimation Methods,* McGraw-Hill, New York.
4. Nirmalakhandan, N.N. and Speece, R.E. (1988), QSAR model for predicting Henry's constant, *Environ. Sci. Technol.,* 22:1349–1357.
5. Suzuki, T. and Kudo, Y. (1990), Automatic log P estimation based on combined additive modeling methods, *J. Comput. Aided Mol. Des.,* 4:155–198.
6. Mackay, D., Paterson, S., Cheung, B., and Nealy, W. (1985), Evaluation of the environmental behavior of chemicals with a level III fugacity model, *Chemosphere,* 14(3/4):335–375.
7. Karickhoff, S.W. (1981), Semi-empirical estimation of sorption of hydrophobic pollutants on natural sediments and soils, *Chemosphere,* 10:833–846.

8. Schwarzenbach, R. and Westall, J. (1981), Transport of nonpolar organic compounds from surface water to groundwater: laboratory sorption studies, *Environ. Sci. Technol.*, 15:1360–1367.

9. Barrow, G.M. (1977), *Physikalische Chemie (Physical Chemistry)*, Vol. 3, 3rd ed. Bohmann, Vienna.

10. Gates, D.M. (1980), *Biophysical Ecology*, Springer, New York.

11. Brüggemann, R., Trapp, S., and Matthies, M. (1991), Behavior assessment of a volatile chemical in the Rhine River, *Environ. Toxicol. Chem.*, 10:1097–1103.

12. Matthies, M., Behrendt, H., and Münzer, B. (1987), EXSOL-Modell für den Transport und Verbleib von Stoffen im Boden (EXSOL-Model for the Transport and Fate of Compounds in Soil), GSF Report 23/87, Munich-Neuherberg, Germany.

13. Trenkle, R. and Münzer, B. (1987), EXAIR Analytisches Transportmodell für die atmosphärische Mischungsschicht (Analytical transport model for the atmospheric mixing layer), UBA Research Report 106 04 016, Appendix I.2, GSF-Neuherberg, Germany.

14. Mackay, D. (1979), Finding fugacity feasible, *Environ. Sci. Technol.*, 13:1218–1223.

15. Netter, H. (1951), *Biologische Physikochemie (Biological Physico Chemistry)*, Akademische Verlagsgesellschaft Athenaion, Potsdam, Germany.

16. Shone, M. and Wood, A. (1974), A comparison of the uptake and translocation of some organic herbicides and a systemic fungicide by barley, I. Absorption in relation to physicochemical properties, *J. Exp. Bot.*, 25:390–400.

17. Mc Farlane, C. and Wickliff, C. (1985), Excised barley root uptake of several 14-C labeled organic compounds, *Environ. Monit. Assess.*, 5:385–391.

18. Briggs, G.G., Bromilow, R.H., and Evans, A.A. (1982), Relationships between lipophilicity and root uptake and translocation of nonionized chemicals by barley, *Pestic. Sci.*, 13:495–504.

19. Trapp, S. and Pussemier, L. (1991), Model calculations and measurements of uptake and translocation of carbamates by bean plants, *Chemosphere*, 22(3-4):327–339.

20. Briggs, G.G., Bromilow, R.H., Evans, A.A., and Williams, M. (1983), Relationships between lipophilicity and the distribution of nonionized chemicals in barley shoots following uptake by the roots, *Pestic. Sci.*, 14:492–500.

21. Kerler, F. and Schönherr, J. (1988), Permeation of lipophilic chemicals across plant cuticles: prediction from partition coefficients and molar volumes, *Arch. Environ. Contam. Toxicol.*, 17:7–12.

22. Campbell, G.S. (1985), *Soil Physics with Basic Transport Models for Soil-Plant Systems*, Elsevier, Amsterdam, The Netherlands.

23. Fußeder, A. (1985), Verteilung des Wurzelsystems von Mais im Hinblick auf die Konkurrenz um Makronährstoffe, (Distribution of the root system of maize considering the competition for macro elements), *Z. Pflanzenernaehr. Bodenkd.*, 148:321–334.

24. Bromilow, R.H. and Chamberlain, K. (1989), Mechanisms and regulation of transport processes. In *Designing Molecules for Systemicity*, British Plant Growth Regulatory Group, Monograph 18, pp. 113–128.

25. Huber, B. (1956), Die Saftströme der Pflanze (The Sap Streams of the Plant), Springer-Verlag, Berlin.

26. Penning de Vries (1975), *Ann. Bot.*, 39:77–92.

27. Bromilow, R.H., Rigitano, R., Briggs, G.G., and Chamberlain, K. (1987), Phloem translocation of nonionized chemicals in *Ricinus communis, Pestic. Sci.,* 19:85–99.
28. Thompson, N. (1983), Diffusion and uptake of chemical vapour volatilising from a sprayed target area, *Pestic. Sci.,* 14:33–39.
29. Möller, F. (1973), *Einführung in die Meteorologie* I. (*Introduction to Meteorology,* Vol. 1), B.I. Hochschultaschen bücher, Bibliographisches Institut AG, Mannheim, Germany.
30. Schönherr, J. and Riederer, M. (1989), Foliar penetration and accumulation of organic chemicals in plant cuticles, *Rev. Environ. Contamin. and Toxicol.,* 108:2-70.
31. Bauer, H. (1991), Mobilität organischer Moleküle in der pflanzlichen Kutikula (Mobility of Organic Molecules in the Plant Cuticle), Thesis, Technical University of Munich, Germany.
32. Langebartels, C., and Harms, H. (1984), Metabolism of pentachlorophenol in suspension cultures of soybean and wheat: pentachlorophenol glucoside formation, *Z. Pflanzenphys.,* 113:201–211.
33. Matthies, M., Behrendt, H., Brüggemann, R., Münzer, B., and Trapp, S. (1992), Exposure and Ecotoxicity for Environmental **CHEM**icals **E4CHEM**. A Simulation Program for the Fate of Chemicals in the Environment, Version 3.1/beta, GSF-München-Neuherberg, Germany.
34. Taylor, A.W. and Glotfelty, D.E. (1988), Evaporation from soils and crops. In *Environmental Chemistry of Herbicides,* Vol. I, R. Grover, Ed., CRC Press, Boca Raton, FL.
35. Mc Farlane, J.C. and Pfleeger, T. (1986), Plant Exposure Laboratory and Chambers, Vol. 1, EPA-600/3–86–007a. U.S. Environmental Protection Agency, Corvallis, OR.
36. Mc Farlane J.C. and Pfleeger, T. (1987), Plant exposure chambers for study of toxic chemical-plant interactions, *J. Environ. Qual.,* 16(4),361–371.
37. Boersma, L., Mc Farlane, J.C., and Lindstrom, F.T. (1991), Mathematical model of plant uptake and translocations of organic chemicals: application to experiments, *J. Environ. Qual.,* 20:137–146.
38. Worthing, C., Ed. (1987), *The Pesticide Manual. A World Compendium,* , 8th ed., Lavenham Press Ltd., Suffolk, U.K.
39. DFG Deutsche Forschungsgemeinschaft (German Research Society), Ed. (1986), *Datensammlung zur Toxikologie der Herbizide (Data to the toxicology of herbicides),* VCH Chemie, Weinheim, Germany.
40. Mc Farlane, J.C., Pfleeger, T., and Fletcher, J. (1990), Effect, uptake and disposition of nitrobenzene in several terrestrial plants, *Environ. Toxicol. Chem.,* 9:513–520.
41. Trapp, S. (1992), Modellierung der Aufnahme anthropogener organischer Chemikalien in Pflanzen (Modeling the uptake of anthropogenic organic chemicals into plants). Doctoral thesis, Technical University of Munich, Germany.
42. Trapp, S., Matthies, M., Scheunert, I., and Topp, E.M. (1990), Modeling the bioconcentration of organic chemicals in plants, *Environ. Sci. Technol.,* 24(8):1246–1252.

43. Jungk, A. (1990), Grundlage für die rationelle Nutzung der Pflanzennährstoffe des Bodens (Base for the rational use of fertilizers in the soil). In *Pflanzenproduktion im Wandel (Plant Production in Change)*, Haug, G., Schuhmann, G., and Fischbeck, G., Eds., VCH, Weinheim, Germany, 197–225.

44. Topp, E.M. (1986), Aufnahme von Umweltchemikalien in die Pflanze in Abhängigkeit von physikalisch-chemischen Stoffeigenschaften (Uptake of Environmental Chemicals into Plants Related to Physicochemical Substance Properties), Doctoral thesis, Technical University of Munich, Germany.

45. Schroll, R. and Scheunert, I. (1992), A laboratory system to determine separately the uptake of organic chemicals from soil by plant roots and by leaves after vaporization, *Chemosphere,* 24(1):97–108.

46. Poigner and Schlatter (1983), Animal toxicology of chlorinated dibenzo-*p*-dioxins, *Chemosphere,* 12:453–462.

47. Gough, M. (1990), Revisiting 1 ppb as a "level of concern" for dioxin in soil, *Organohalogen Compounds,* 3:323–326.

48. BMFT/GSF Workshop, Ursachen und quantitative Beziehungen bezüglich der Belastung von Futter- und Nahrungspflanzen sowie bodengehaltener Nutztiere mit Dioxinen (Reasons for and Quantitative Relations of the Contamination of Plants and Animals with Dioxins). Munich/Neuherberg, Germany, April 22–23, 1992.

49. Dioxin-Informationsveranstaltung (Dioxin Information Arrangement), Augsburg, Germany, November 11–13, 1991.

50. McLachlan, M., University of Bayreuth, Germany, personal communication, 1992.

51. Kjeller, L.-O., Jones, K.C., Johnston, A.E., and Rappe, C. (1991), Increases in the polychlorinated dibenzo-*p*-dioxin and -furan content of soils and vegetation since the 1840s, *Environ. Sci. Technol.,* 25(9):1619–1627.

52. McLachlan, M. (1992), Das Verhalten hydrophober chlororganischer Verbindungen in laktierenden Rindern (Behavior of Hydrophobic Chloroorganic Compounds in Lactating Cows), Doctoral thesis, University of Bayreuth, Germany.

53. Kew, G.A., Schaum, J.L., White, P., and Evans, T.T. (1989), Review of plant uptake of 2,3,7,8-TCDD from soil and potential influences of bioavailability, *Chemosphere,* 18:1313–1318.

54. McCrady, J., Mc Farlane, J.C., and Gander, L. (1990), The transport and fate of 2,3,7,8-TCDD in soybean and corn, *Chemosphere,* 21(3):359–376.

55. Rippen, G. (1991), *Handbuch Umweltchemikalien (Handbook of Environmental Chemicals).* ecomed, Landsberg am Lech, Germany.

56. McCrady, J.K. (1993) Uptake and photodegradation of 2,3,7,8-tetrachlorodibenzo-*p*-dioxin sorbed to grass foliage, *Environ. Sci. Technol.,* 27(2): 343–350.

CHAPTER **6**

Partitioning and Transport of Organic Chemicals between the Atmospheric Environment and Leaves

Markus Riederer

TABLE OF CONTENTS

1-56670-078-7/95/$0.00+$.50
© 1995 by CRC Press, Inc.

I. INTRODUCTION

Leaves are the sites of the most intimate contact between terrestrial plants and their atmospheric environment. Indeed, plants can draw considerable selective advantages from harnessing as much solar radiation as possible and from facilitating the uptake of atmospheric CO_2. Both aims can be achieved by developing large surface areas. A fully grown tree may thus bear several hundred thousand leaves, and the combined leaf area of stands of forest trees or crop plants may exceed up to 20 times the area they are growing on.[1,2]

When plants have to live under the conditions of a technical civilization, the air and the different forms of deposition (precipitation, particulate matter) their leaves contain, in addition to the natural constituents, a complex mixture of organic contaminants originating from industrial activity, combustion processes in traffic and homes, and from agriculture. Therefore, questions arise with regard to (1) how these chemicals may interact with the leaves of plants, (2) how and to what degree they are taken up into the interior of the leaves, and (3) what role plants consequently play in passing these chemicals into terrestrial food chains. These are the basic questions stimulating the interest in the atmospheric environment-to-leaf transfer of organic chemicals.[5]

At the present state of knowledge, these questions cannot be answered yet by formulating a comprehensive model of leaf/atmospheric environment exchange of organics. What can be done, however, is to identify the key processes involved and to explore how these processes are affected by the physicochemical properties of the chemicals and the properties of the plants.

Hence, this chapter will deal with partitioning between the different compartments of the leaf/atmospheric environment system and with the diffusive transport phenomena occurring at the interface and in the interior of the leaf. Whenever possible, this treatment will be quantitative. Therefore, the necessary equations and relationships will be presented explicitly and some applications will be demonstrated using a set of reference chemicals of diverse physico-chemical properties.

At the same time, the basic biological information relevant to leaf/atmosphere exchange processes will be provided. Plants are highly organized living systems with complex and diverse structural and compositional properties. It would be surprising if this complexity and diversity of properties and organization did not affect or even decisively determine the uptake of organic compounds from the atmospheric environment. Consequently, a further objective of this chapter is to draw attention to these biological aspects sometimes neglected in the past.

As a whole, the intention of this chapter is to outline the basic botanical and physicochemical concepts necessary for understanding and predicting the behavior of organic compounds in the atmospheric environment/leaf system. It is the conviction of the author that the whole field is served better by developing a sound theoretical basis than by a review of the experimental work done so far on this subject.

II. THE PARTITIONING OF ORGANIC CHEMICALS BETWEEN THE ATMOSPHERIC ENVIRONMENT AND PLANT LEAVES

The simplest way of treating atmospheric environment/foliage exchange of organic chemicals is the equilibrium view. This approach assumes that, within the time-frame chosen, (1) environmental conditions remain stable, (2) the leaf does not grow, (3) the compound is not metabolized within the plant, and (4) the transport processes proceed fast enough to allow complete equilibration. The equilibrium view of atmospheric environment/foliage partitioning helps to outline the basic principles of this process despite its obvious inherent limitations (see Section III for further discussion). It can, however, provide useful information, for instance, on the main sorption compartments of the leaf, on the partitioning between these compartments, on the maximum atmospheric input, and on the relative importance of the different leaf compartments. The objective of the present section is to give simple and, whenever possible, quantitative answers to such problems.

A. THE COMPARTMENTS OF THE ATMOSPHERIC ENVIRONMENT/LEAF SYSTEM

The system we are dealing with is the plant leaf (or needle) and its surrounding atmospheric environment. This can be considered as a microscopic replication of larger-scale systems like those used for studying the global distribution of chemicals between, for instance, the atmosphere, seawater, sediment, and fish. In contrast to such systems, however, the leaf has a much greater complexity of composition and spatial order (see also Section III.A).

When the partitioning of lipophilic organic compounds is considered, the various constituents of plant biomass can be grouped according to their overall properties (Table 1). The total lipid content of a leaf is made up of two fractions, the internal and the surface lipids. This separation is justified because of their spatial arrangement in the system, which will become relevant when the kinetics of the partitioning process is treated (see Section III). Another reason is the difference in the physical states of both fractions: while internal lipids are predominantly *liquid,* surface lipids are either polymeric (cutin) or more or less crystalline *solids* (cuticular waxes).

B. PARTITIONING BETWEEN INDIVIDUAL COMPARTMENTS OF THE ATMOSPHERIC ENVIRONMENT/LEAF SYSTEM

1. Definition of Partition Coefficients

The first step in describing and analyzing equilibrium partitioning in a multicompartment system is to establish the relationships between the different compartments. The ratios between the equilibrium concentrations C (in molar, i.e., volume-related units) in any two compartments i and j are called the partitioning coefficients:

$$K_{ij} = \frac{C_i}{C_j} \tag{1}$$

Accordingly, for instance, the equilibrium distribution of a chemical between either the air or the cuticle phase and the aqueous compartment can be expressed by the air/water and cuticle/water partition coefficients,

$$K_{AW} = \frac{C_A}{C_W} \tag{2}$$

and

$$K_{CW} = \frac{C_C}{C_W} \tag{3}$$

respectively.

It is not necessary to determine experimentally the values of the whole set of partition coefficients between all compartments of a system. Rather, the system can be divided into three-phase subsystems consisting of phases $i, j,$ and k. Then, the partition coefficient K_{ik} can be estimated from

$$K_{ik} = \frac{K_{ij}}{K_{kj}} = \frac{C_i \cdot C_j}{C_j \cdot C_k} \tag{4}$$

Table 1. Compartments of leaf/atmosphere system and its constituents

Compartment	Constituents	Subscript
Air	Atmosphere	A
	Intercellular air	
Water	Precipitation	W
	Interstices of cell walls	
	Cytoplasm	
	Vacuole	
	Xylem tissue	
Polar	Polysaccharides	K
	Inorganic salts	
	Most of organic contents of cytoplasm and vacuole	
Internal lipids	Membrane lipids	G
	Storage lipids	
	Resins	
	Essential oils	
Surface lipids	Cuticle (consisting of cutin and waxes)	C

when K_{ij} and K_{kj} are known. Thus, for example, the cuticle/air partition coefficient, K_{CA}, can be derived from

$$K_{CA} = \frac{K_{CW}}{K_{AW}}$$

(5)

2. Values of Partition Coefficients

Experimental values of partition coefficients are available for the distribution of lipophilic organics between the relevant compartments of the atmospheric environment/leaf system (Table 1). These are the air (subscript A), water (subscript W), internal lipid (subscript G), and surface lipid or cuticle (subscript C) compartments. In a first approximation, the polar compartment (subscript K) can be neglected when studying the partitioning of lipophilic compounds.

Due to its abundance and its ubiquity, the aqueous compartment can be comfortably chosen as a reference phase to which the equilibrium concentrations in the remaining concentrations can be related according to Equation 4. Thus, partitioning within the whole system is completely described by the air/water (K_{AW}), the internal lipid/water (K_{GW}), and the cuticle/water (K_{CW}) partition coefficients.

The former two partition coefficients can be easily derived from fundamental physicochemical properties of the compounds in question. The air/water partition coefficient is related to the Henry's law constant H by the equation $K_{AW} = H/(R \cdot T)$, and the internal lipid/water partition coefficient is close to the 1-octanol/water partition coefficient according to the equation $K_{GW} \approx K_{OW}$.[3,4] The cuticle/water partition coefficient remains; fortunately, it can be determined using isolated cuticles.[6,7]

Experimental values of K_{CW} have been determined for a range of organic solutes of widely varying chemical structures and physicochemical properties.[6-10] *Citrus aurantium* L. leaf cuticle/water partition coefficients measured so far range from 0.078 for methanol to 1.66×10^7 for bis(2-ethylhexyl)phthalate (DEHP; Table 2). Partition coefficients determined for *Capsicum annuum* L. fruit cuticles are somewhat higher but always remain in the same order of magnitude (Table 2). The values of cuticle/water partition coefficients for 2,4-(dichlorophenoxy)acetic acid (2,4-D) varied more within one species than between 11 plant species.[6] This fact must be kept in mind whenever the accumulation of organic chemicals in "vegetation" is to be modeled. There is no cuticle from one plant species with properties representative for all others, and there is even a large degree of intraspecific variability (supposedly due to genetic, environmental, and developmental factors).

With highly volatile compounds like organic solvents experimental difficulties arise when K_{CW} is to be determined. Consequently, the cuticle/air partition coefficients of these compounds are determined instead,[11] and the corresponding values of K_{CW} are estimated according to Equation 5 (Table 3).

In equilibrium partitioning, the polymeric cutin matrix of plant cuticles is the main compartment for the sorption of lipophilic chemicals. At low concentrations of sorbate in the cuticle, comparable to those resulting from environmental contamination, the sorptive properties of the plant cuticle resemble those of a slightly polar aliphatic solvent like 1-octanol.[12]

The partitioning of weak electrolytes between an aqueous and a lipid phase is affected by the pH of the aqueous solution since it is the nondissociated species of the electrolyte that is lipophilic enough to partition into the lipid compartment.[3] For instance, in the cuticle/water system, the actual partition coefficient K_{CW}^{pH} at a given pH can be derived from the K_{CW} of the nondissociated species (like those tabulated in Table 2) by

$$K_{CW}^{pH} = \frac{K_{CW}}{1 + 10^{pH - pK_a'}} \tag{6}$$

where pK_a' is the acid dissociation constant of the compound corrected for ionic strength. In the case of weak bases the appropriate pK_a' is obtained from tabulated values of basic dissociation constants pK_b' according to the equation $pK_a' = 14 - pK_b'$.

Another complication which also may be relevant under an environmental point of view is the temperature dependence of the partition coefficients. The variation of the partition coefficient K_{ij} with temperature is given by the general relationship for the temperature dependence of any equilibrium constant:

$$\ln \frac{K_{ij}^1}{K_{ij}^2} = \frac{\Delta H^\circ}{R} \left(\frac{1}{T^2} - \frac{1}{T^1} \right) \tag{7}$$

where K_{ij}^1 and K_{ij}^2 are the partition coefficients at temperatures (in K) T^1 and T^2, respectively. ΔH° stands for the change in the standard partial molar

Table 2.   Cuticle/water partition coefficients for *Citrus aurantium* leaf and *Capsicum annuum* fruit cuticles at 25°C[a]

Common name	Compound	$\log K_{cw}$ Citrus leaf	$\log K_{cw}$ Capsicum fruit
Methanol	Methanol	−1.11	—
Primicarb	5,6-Dimethyl-2-dimethylamino-4-pyrimidinyl-dimethylcarbamate	—	1.20
Phenol	Phenol	—	1.59
2-NP	2-Nitrophenol	—	1.92
4-NP	4-Nitrophenol	1.79	1.97
Atrazine	6-Chloro-N-ethyl-N'-(1-methylethyl)-1,3,5-triazin-2,4-diamine	2.15	2.19
1-NAA	1-Naphthaleneacetic acid	2.18	2.33
2,4-D	2,4-(Dichlorophenoxy)acetic acid	2.47	2.76
Bentazon	3-(1-Methylethyl)1H-2,1,3-benzothiadiazin-4(3H)-on-2,2-dioxide	—	2.78
Birlane	Diethyl-[2-chloro-1-(2,4-dichlorophenyl)]-vinyl-phosphate	—	3.20
2,4,5-T	2,4,5-(Trichlorophenoxy)acetic acid	3.13	3.21
Triadimenol	β-(4-Chlorophenoxy)-α-(1,1-dimethylethyl)-1H-1,2,4-triazole-1-ethanol	3.37	3.37
Tebuconazole	α-2-(4-Chlorophenyl)-ethyl-α-1,1-dimethylethyl)-1H-1,2,4-triazole-1-ethanol	—	3.54
WL 110547	1-(3-Fluoromethylphenyl)-5-phenoxy-1,2,3,4-tetrazole	—	3.60
Bitertanol	β-([1,1'-Biphenyl]-4-yloxy)-α-1,1-dimethylethyl)1H-1,2,4-triazole-1-ethanol	3.77	3.85
PCP	Pentachlorophenol	4.42	4.66
HCB	Hexachlorobenzene	5.70	5.80
Perylene	Perylene	6.45	6.55
DEHP	bis(2-Ethylhexyl)phthalate	7.22	7.48

[a] Data compiled from References 6 through 10; when appropriate, K_{cw} are given for the nondissociated species.

enthalpy of the phase transfer process. For the cuticle/water system, the thermodynamics of the partitioning of 4-nitrophenol (4-NP) has been studied.[12] At low concentrations of solute in the cuticle phase, $\Delta H°$ is about −24 kJ · mol^{-1}. The negative sign indicates that the transfer of an organic molecule from its aqueous solution to the lipophilic cuticle is an exothermic process. Thus, increasing temperature will shift the equilibrium toward the aqueous phase and, consequently, partition coefficients will decrease.

For instance, the K_{cw} of 4-NP at 10°C (a representative mean annual temperature in temperature climates) will be 1.67 times higher than the value determined at 25°C. The amount of heat released upon transfer of a molecule from one phase to another is greater the more intensive its interactions with the new environment are. In the case of the partitioning of organic compounds between aqueous solutions and lipid phases, $\Delta H°$ increases with lipophilicity. Therefore, the temperature dependence of K_{cw} can be expected to become more pronounced with decreasing polarity.

Table 3.   Cuticle/air and cuticle/water partition coefficients at 25°C for volatile organic compounds[a]

Compound	log K_{CA}	log K_{CW}[b]
Methanol	2.63	−1.07
Ethylmethylketone	2.56	−0.03
1-Butanol	3.56	0.39
Ethylacetate	2.53	0.61
1-Nitropropane	3.39	0.79
n-Butylacetate	3.39	1.64
Benzene	2.63	2.03
Isoprene	1.59	2.08
Tetrachloromethane	2.43	2.33
1,1,2-Trichloroethene	2.88	2.56
1,1,1-Trichloroethane	2.54	2.58
Toluene	3.16	2.61
Chlorobenzene	3.52	2.70
o-Xylene	3.56	2.80
Ethylbenzene	3.40	2.91
Cyclohexane	2.22	3.08
Limonene	4.04	4.26
n-Heptane	2.50	4.47

[a] Data from Reference 11; isolated and solvent-extracted *Lycopersicon esculentum* fruit cuticles were used for experiments.

[b] Estimated according to $K_{CW} = K_{CA} \times K_{AW}$.

3. Estimating Cuticle/Water Partition Coefficients

Several ways to predict quantitatively the values of K_{CW} from fundamental properties or molecular structure have been explored. This is a necessary and rewarding objective in view of the very large number of biogenic and anthropogenic organic compounds which may come into contact with leaf surfaces. A quantitative property-property relationship (QPPR) was established between the cuticle/water partition coefficient and the 1-octanol/water partition coefficient:[13]

$$\log K_{CW} = 0.057 + 0.970 \log K_{OW} \ (r = 0.987) \tag{8}$$

A similar correlation but with a slightly lower degree of determination was obtained between K_{CW} and the solubility of the compound in water (S_W, mol · l^{-1}) at 25°C:[13]

$$\log K_{CW} = 1.118 - 0.569 \log S_W \ (r = 0.978) \tag{9}$$

Reliable data on K_{OW}, or aqueous solubility, are available only for a limited number of compounds. Additionally, relationships between two empirically determined properties like those in Equation 8 and 9 are, for fundamental statistical reasons, prone to large predictive errors.[9] Therefore, a quantitative structure-property relationship (QSPR) was established between K_{CW} and nonempirical descriptors of molecular structure:[9]

$$\log K_{CW} = 0.37 + 1.31\,^3\chi^v - 1.49 N_{aliph}^{OH}\,(r = 0.992) \qquad (10)$$

The descriptors used are the valence third-order connectivity index ($^3\chi^v$) calculated from the connectivity matrix of the molecules (for further information on the connectivity index concept see Reference 14) and, as an *ad hoc* descriptor, the number of hydroxy groups attached to aliphatic chains (N_{aliph}^{OH}). The advantage of this approach is that it allows the estimation of values of K_{CW} even for compounds for which only the structural formulas are known. Recent research, however, has shown that the predictive power of Equation 10 is restricted to relatively large molecules with cyclic constituent groups, like those listed in Table 2.[15] Work is currently in progress to include also small and straight-chain molecules into a nonempirical predictive tool similar to Equation 10.

C. THE ATMOSPHERE/LEAF PARTITIONING OF VOLATILE ORGANIC COMPOUNDS

In the preceding parts of this section, partitioning between different compartments of the atmospheric environment/leaf system has been described. For this purpose, only the material nature of the different phases had to be taken into account. As a consequence, the results were restricted to simple partitioning between individual phases, whereas a treatment of the distribution of a chemical between the outside environment and the leaf as a whole was not possible.

To achieve this goal the leaf is regarded as a mixed phase composed of thoroughly intermingled materials belonging to the different compartments outlined in Table 1. The overall properties of this mixture of phases are a function of the volume ratios of the different compartments and their respective partition coefficients.

In quantitative terms, a leaf/air partition coefficient can be defined by

$$K_{LA} = v_A + \frac{v_W}{K_{AW}} + v_C K_{CA} + v_G K_{GA} \qquad (11)$$

where subscripts describe phases according to their usual meaning (Table 1) and v_i stand for the volume fractions of the individual compartments relative to the volume of the leaf ($v_i = V_i/V_L$).[16]

This is the moment when the fairly abstract chemical engineering-like treatment of partitioning meets its limitations as the real plant and its properties come into play. Now the dimensions, anatomy, and composition of the leaf are as important as the values of the partition coefficients themselves because these plant-related properties determine the size of the volume fractions (v_i). In the following, the leaves of European beech (*Fagus sylvatica* L.) will be used as representative of the foliage of deciduous tree species in temperate climates. A comprehensive set of data on the dimensions, properties, and composition of these leaves is available from data in the literature[17-21] or can be deduced from general botanical[22] and plant physiological knowledge[23] (Table 4).

Using a set of reference chemicals of diverse physicochemical properties (Table 5, References 24 and 25) and experimentally determined values of K_{CW} (Table 2), the range and variability of K_{LA} can be explored (Table 6). The leaf-air partition coefficients (expressed on a volume basis) range from 8.5 for toluene to 7.1×10^7 for perylene. From Equation 11 it is clear that high lipid/water and low air/water partition coefficients will increase the concentration of a chemical in the leaf relative to its vapor-phase concentration in the surrounding atmosphere.

The effect of lipophilicity and volatility from water on atmosphere/leaf partitioning can be further studied by calculating K_{LA} according to Equation 11 for compounds of varying K_{OW} and K_{AW} (Figure 1). For this purpose, it is assumed that K_{CW} for *F. sylvatica* leaves is related to K_{OW} as given by Equation 8 over the whole range of octanol/water partition coefficients. The resulting family of curves shows that K_{LA} is inversely proportional to K_{AW} and that it increases with K_{OW}. An interesting feature of Figure 1 is the fact that, for a given K_{AW}, the leaf/atmosphere partition coefficients remain constant for values of K_{OW} up to log $K_{OW} \approx 2.5$. Within this range of K_{OW} leaf/atmosphere partitioning is predominated by the dissolution of the chemical in the large aqueous compartment of the leaf. The contribution of the lipid compartments $(v_C K_{CA} + v_G K_{GA})$ to K_{LA} is negligible. With increasing K_{OW}, $(v_C K_{CA} + v_G K_{GA}) \ll v_W/K_{AW}$ no longer holds, for compounds with log $K_{OW} \geq 3$ and a given K_{AW}, log K_{LA} increases linearly with log K_{OW}.

The value of K_{OW} where the transition from aqueous-dissolution- to lipid-partitioning-dominated accumulation of organic compounds in leaves occurs depends on the composition and the relative size of the aqueous, internal lipid, and surface lipid compartments and is likely to vary among species.

The analysis of leaf/atmosphere partitioning also shows that the intercellular air space making up the second largest volume fraction ($v_A \approx 0.3$ in most species) contributes to the total concentration of a chemical in the leaf only in the case of very volatile compounds. Equation 11 indicates that intercellular air will become important only when the relationship $v_A \ll v_W/K_{AW} + v_C K_{CA} + v_G K_{GA}$ does not hold any longer.

Leaf/atmosphere partition coefficients can be used to estimate the maximum (i.e., equilibrium) contamination of foliage from vapor-phase air pollution. For typical concentrations in the atmosphere contamination levels (on a dry weight basis) in the ppb range are predicted for toluene and pentachlorophenol (PCP), while the concentration of bis(2-ethylhexyl)phthalate (DEHP) will be on the order of ppt (Table 7). When information is available not only on the dimensions and composition of the leaves, but also on the standing leaf

Table 4. Selected properties of model leaf based on those of European Beech (*Fagus sylvatica* L.)

Leaf dimensions		
Total leaf area (A_L)	1.0×10^{-2} m^2	Top + bottom surfaces; Ref. 17
Leaf thickness (x_L)	2.0×10^{-4} m	Ref. 18
Leaf volume (V_L)	1.0×10^{-6} m^3	$V_L = A_L/2 \cdot x_L$
Cuticle		
Cutin coverage	6.3×10^{-4} kg $\cdot$ m^{-2}	Related to A_L; Ref. 19
Wax coverage	9.0×10^{-5} kg $\cdot$ m^{-2}	Related to A_L; Ref. 17
Thickness of cuticle	5.7×10^{-7} m	From cutin coverage
Volume fractions		
Intercellular air (v_A)	0.300	
H$_2$O (v_W)	0.643	
Polar substances (v_K)	0.052	
Lipid (v_G)	0.002	
Cuticle (v_C)	0.003	
Densities		
H$_2$O (ρ_W)	1000 kg $\cdot$ m^{-3}	
Polar (ρ_K)	1300 kg $\cdot$ m^{-3}	
Lipid (ρ_G)	800 kg $\cdot$ m^{-3}	
Cuticle (ρ_C)	1100 kg $\cdot$ m^{-3}	Ref. 20
Fresh leaf (ρ_L^{fw})	716 kg $\cdot$ m^{-3}	$\Sigma\ v_i\rho_i$
Dry leaf (ρ_L^{dw})	72.5 kg $\cdot$ m^{-3}	$\Sigma\ v_i\rho_i - v_W\rho_W$
Composition of leaf		
Leaf fresh weight	1.43×10^{-1} kg $\cdot$ m^{-2}	$\rho_L^{fw} x_L$; related to A_L
Leaf dry weight	1.45×10^{-2} kg $\cdot$ m^{-2}	$\rho_L^{dw} x_L$; related to A_L
H$_2$O content	899 g $\cdot$ kg^{-1} FW	
Polar substances content	94.5 g $\cdot$ kg^{-1} FW	932 g $\cdot$ kg^{-1} DW
Lipid content	2.24 g $\cdot$ kg^{-1} FW	22.1 g $\cdot$ kg^{-1} DW
Cuticle content	4.61 g $\cdot$ kg^{-1} FW	13.8 g $\cdot$ kg^{-1} DW
Stomata and boundary layer		
Relative pore area (na_{st})	3.0×10^{-3} m^2 $\cdot$ m^{-2}	Ref. 23
Effective pore depth (x_{st})	5.0×10^{-6} m	Ref. 23
Mean pore radius (y_{st})	5.0×10^{-6} m	Ref. 23
Thickness of boundary layer (x_{st})	1.0×10^{-3} m	Wind speed = 0.8 m $\cdot$ s^{-1}
Stand data		
Leaf production (dry weight)	3.3×10^3 kg $\cdot$ ha^{-1} $\cdot$ a^{-1}	Ref. 21
Leaf production (fresh weight)	3.3×10^4 kg $\cdot$ ha^{-1} $\cdot$ a^{-1}	Using $\rho_L^{fw}/\rho_L^{dw} = 9.88$
Total leaf surface area	2.3×10^5 m^2 $\cdot$ ha^{-1}	Resulting leaf area index =11.4
Total amount of cutin	145 kg $\cdot$ ha^{-1}	
Total amount of wax	20.7 kg $\cdot$ ha^{-1}	
Total amount of lipids	73.9 kg $\cdot$ ha^{-1}	

Note: FW = fresh weight; DW = dry weight.

biomass of a given type of vegetation (Table 4), then the foliage-mediated input from the vapor phase can be assessed. In the case of *F. sylvatica* stands, typical concentrations in the air[24,26] lead to inputs on the order of some mg $\cdot$ ha^{-1} for toluene and PCP (Table 7). Extreme values can be estimated for DEHP (ca. 1 g $\cdot$ ha^{-1} $\cdot$ a^{-1}) and 2,3,7,8-tetrachlorodibenzodioxin (TCDD, 4.4×10^{-7} g $\cdot$ ha^{-1} $\cdot$ a^{-1}).

The actual input via the foliage may well be much higher since wet and particulate deposition also contribute considerably to the load of organic

Table 5. Physicochemical properties of reference compounds[a]

Compound	MW (g · mol⁻¹)	p^s (Pa)	S_W (mol · m⁻³)	log K_{OW}	log K_{AW}
Methanol	32	$1.52 \times 10^{+4}$	—	−0.71	−3.85
Phenol	94	$8.30 \times 10^{+1}$	$8.71 \times 10^{+2}$	1.39	−4.42
2-NP	139	$1.19 \times 10^{+1}$	$9.35 \times 10^{+0}$	1.69	−3.29
4-NP	139	5.40×10^{-3}	$1.06 \times 10^{+2}$	1.92	−7.69
2,4-D	221	$1.00 \times 10^{+0}$	$1.81 \times 10^{+0}$	2.50	−3.65
Toluene	92	$3.85 \times 10^{+3}$	$5.75 \times 10^{+0}$	2.62	−0.57
Atrazine	216	4.00×10^{-5}	1.39×10^{-1}	2.64	−6.94
2,4,5-T	256	5.00×10^{-3}	8.61×10^{-1}	3.40	−5.63
PCP	266	2.00×10^{-3}	4.50×10^{-2}	4.07	−4.75
HCB	285	1.00×10^{-3}	1.40×10^{-4}	5.47	−2.54
Perylene	252	7.00×10^{-7}	1.60×10^{-6}	6.50	−3.75
DEHP	391	1.10×10^{-6}	5.90×10^{-8}	7.86	−2.12

[a] At 25°C, all partition coefficients are on a volume-to-volume basis; saturation vapor pressures (p^s) and aqueous solubilities (S_W) from References 24 and 25. 1-octanol/water partition coefficients (K_{OW}) from Reference 7, air/water partition coefficients (K_{AW}) estimated according to $K_{AW} = p^s/(S_W RT)$. MW = molecular weight. See Table 2 for compound formulas.

Table 6. Leaf/air partition coefficients, K_{LA}, and distribution of the total amount of a chemical between individual leaf compartments, M_i/M_L[a]

Compound	log K_{LA}	M_C/M_L	M_W/M_L
Methanol	3.66	0.000	0.999
Phenol	4.31	0.092	0.844
2-NP	3.24	0.169	0.722
4-NP	7.67	0.142	0.682
2,4-D	3.94	0.336	0.336
Toluene	0.93	0.322	0.281
Atrazine	7.20	0.169	0.354
2,4,5-T	6.57	0.343	0.075
PCP	6.66	0.705	0.008
HCB	5.77	0.650	0.000
Perylene	7.85	0.494	0.000
DEHP	7.38	0.201	0.000

[a] Where appropriate the nondissociated species is regarded; subscripts $_C$ and $_W$ denote the cuticular and aqueous compartments of the leaf, respectively. $M_C/M_L + M_W/M_L < 1.000$ since, depending on the compound, M_G and/or $M_A > 0$. See Table 2 for compound formulas.

chemicals in plant leaves (see Section III.E). At this point, it also should be remembered that atmosphere-to-foliage partitioning is a nonspecific process where the distribution of any chemical proceeds independently from that of other compounds present. Thus, the partitioning of organic chemicals between the atmosphere and leaves is cumulative and, given the large number of individual compounds present in the atmospheric environment, a fairly high total burden of organic contaminants in leaves may result.

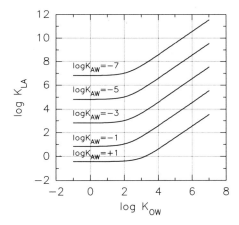

FIGURE 1. Leaf/air partition coefficients (K_{LA}) of organic compounds as functions of the 1-octanol/water (K_{OW}) and air/water (K_{AW}) partition coefficients. The curves were calculated according to Equation 11 using the leaf properties described in Table 4.

Table 7. Estimated leaf concentrations and total amounts in the foliage of a *Fagus sylvatica* stand at given vapor-phase concentration in the atmosphere[a]

Compound	C_A[a] (g · m^{-3})	C_L (g · kg^{-1} FW)	C_L (g · kg^{-1} DW)	Amount (g · ha^{-1})
Toluene	6.00×10^{-5}	7.09×10^{-7}	5.67×10^{-6}	2.34×10^{-3}
PCP	1.00×10^{-10}	6.35×10^{-7}	5.08×10^{-6}	2.09×10^{-3}
DEHP	1.00×10^{-8}	3.33×10^{-4}	2.67×10^{-3}	$1.10 \times 10^{+0}$
TCDD	2.70×10^{-15}	1.34×10^{-10}	1.07×10^{-9}	4.43×10^{-7}

[a] Representative vapor-phase concentrations (C_A) taken from References 24 and 26 (for 2,3,7,8-tetrachlorodibenzodioxin, TCDD). See Table 2 for other compound formulas. FW = fresh weight; DW = dry weight.

D. THE RELATIVE DISTRIBUTION OF AN ORGANIC CHEMICAL BETWEEN LEAF COMPARTMENTS

Information on the distribution of an organic chemical among the leaf compartments or on the total concentration within the leaf arising from a given vapor-phase background concentration may not be sufficient for certain applications of leaf/atmosphere models. For instance, studies on the vegetation-mediated food-chain input of organic chemicals might require specific information on the preferred accumulation sites of a given compound within the leaf biomass. This means that the relative distribution of the total amount of a chemical contained in the leaf among the different compartments should be known. Examples of the relevance of the accumulation site for determining the extent of human intake of the contaminant are storage lipids used as vegetable oils and fats or skins of fruit which may be peeled.

For this purpose, a fugacity-based model has been devised[16] which is an adaptation from similar large-scale models.[27-30] Basically, this concept relies on fugacities (*escaping tendencies*) instead of concentrations. This elegantly simplifies the treatment of many equilibrium partitioning problems since, at equilibrium, the individual fugacities of a compound in the different compartments, f_i, do not differ and are equal to an overall fugacity f (dimension Pa). In the case of uptake from the vapor phase, f is equal to the vapor pressure of the compound in the atmosphere.

At low concentrations, fugacities are linearly related to concentrations in the different compartments by

$$C_i = Z_i f \tag{12}$$

where Z_i stands for the fugacity capacity (in mol $\cdot$ m^{-3} $\cdot$ Pa^{-1}) of the compartment i. The corresponding capacities of the compartments of the leaf/atmospheric environment system can be calculated from fundamental properties[16,19] according to

$$Z_A = \frac{1}{RT} \tag{13}$$

$$Z_W = \frac{1}{H} \tag{14}$$

$$Z_C = Z_W K_{CW} \tag{15}$$

$$Z_G = Z_W K_{GW} \approx Z_W K_{OW} \tag{16}$$

The amount of a given chemical in one of the leaf compartments, M_i, relative to the total amount in the leaf, M_L, can now be estimated by

$$\frac{M_i}{M_L} = \frac{v_i Z_i}{Z_L} \tag{17}$$

where Z_L is the overall fugacity capacity for the whole leaf:

$$Z_L = \sum_{i=1}^{n} v_i Z_i \tag{18}$$

Applying this concept to *F. sylvatica* leaves and the reference chemicals (Table 5), the relative importance of the aqueous and the cuticular compartments, can be studied (Table 6). With decreasing polarity of the compounds, the proportion of the total amount sorbed in the cuticle increases from zero with methanol to a maximum of 71% with PCP. On the other hand, the aqueous compartment contains most of the more polar compounds like methanol, phenol, 2-nitrophenol (2-NP), and 4-NP, while, despite its large size, it plays no role during the atmospheric environment/leaf partitioning of highly lipophilic compounds like PCP, hexachlorobenzene (HCB), perylene, or DEHP (Table 6).

The relationships between lipophilicity and volatility from aqueous solution on the one hand and M_C/M_L and M_W/M_L on the other can be examined when one assumes again that Equation 8 correctly describes the variation of K_{CW} with K_{OW}. Both ratios are drastically affected by lipophilicity in the range of log K_{OW} from approximately 1 to 4 (Figure 2). The relative amount contained in the cuticle is practically zero for log $K_{OW} \leq 1$ and reaches a maximum of about 45% for log $K_{OW} \approx 4$ (Figure 2A). Volatility affects the relative distribution between the cuticle phase and the rest of the leaf only in the range $0 \leq \log K_{OW} \leq 5$ (Figure 2A). The impact of K_{AW} on the relative distribution between the cuticle and the rest of the *F. sylvatica* leaf is restricted to $-3 \leq \log K_{AW} \leq +1$ and even in this range is comparable small. The relative amounts in the cuticle of a compound with a K_{OW} value of 1000 are 25% for log $K_{AW} = 1$ and 42% for log $K_{AW} \leq 3$.

The relative amount in the aqueous phase is influenced by the lipophilicity of the compound only for log $K_{OW} \leq 4$. Above this value, M_W/M_L is negligible (Figure 2B). However, at values below log $K_{OW} \approx 4$, M_W/M_L is strongly affected by K_{AW}. For volatile compounds ($K_{AW} = 10$), the fraction contained in the aqueous leaf compartment amounts to 20% and decreases to zero in the range $2 \leq \log K_{OW} \leq 4$. Relatively polar ($K_{OW} \leq 10$) and involatile compounds accumulate almost exclusively in the aqueous phase (95% for log $K_{AW} = -1$ and 100% for log $K_{AW} \leq -3$). The importance of the aqueous compartment and the influence of K_{AW} rapidly decline with increasing lipophilicity.

When analyzing the relative distribution of a compound in this way one has to bear in mind that, according to Equation 18, the absolute values of M_i/M_L are directly proportional to the volume fraction v_i of the compartment under consideration. Thus, the data given in Figure 2 are valid only for leaves with properties similar to those of *F. sylvatica* (Table 4). This, again, emphasizes the important role biological complexity and diversity plays in determining atmospheric environment/leaf partitioning of organic compounds.

III. THE KINETICS OF THE UPTAKE OF ORGANIC COMPOUNDS INTO LEAVES

The equilibrium view presented so far (Section II) is instructive and may serve as a first approximation to the real phenomena. Its main merit is that it helps to identify the factors influencing atmospheric environment/leaf partitioning and the main accumulation sites within the leaves. Nevertheless, the predictions derived from equilibrium models of the distribution of organic chemicals between the air and foliage may, to an unknown degree, deviate from reality. This may lead to wrong conclusions concerning the role of

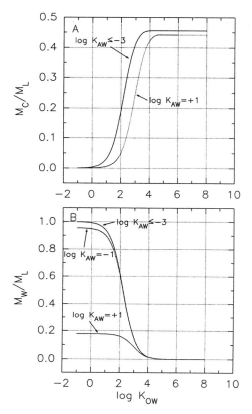

FIGURE 2. The fractions of the total amount of an organic chemical in the leaf contained at equilibrium in (A) the cuticle or (B) the aqueous compartments as functions of the 1-octanol/water (K_{OW}) and air/water (K_{AW}) partition coefficients. The M_i/M_L ratios were calculated according to Equation 17.

vegetation in the fate of pollutants in the environment or the food-chain input of organic chemicals.

The reasons for those shortcomings at least partially lie in the simplifying assumptions made. The equilibrium approach is macroscopically static and thus does not consider the irreversible processes occurring in the nonequilibrium phases of atmospheric environment/leaf exchange. From a kinetic point of view the most questionable assumption implicit in the equilibrium treatment is that equilibration times are negligible within the time frame chosen. In detail this means (1) that all compartments are equally accessible to the chemicals in the atmospheric environment and (2) that the transport properties of the different compartments are comparable. It will be the scope of this section to show under which conditions these assumptions may be violated and how difficulties arising from this fact may be overcome by a kinetic treatment of atmospheric environment/leaf exchange.

A. ACCESSIBILITY OF LEAF COMPARTMENTS

Leaves, like any other plant organ, share one basic feature with other living organisms: they are complex in terms of structure, i.e., the spatial organization of compartments and their constituents. Compartments are not thoroughly mixed, rather they are organized to form strictly defined cytological and anatomical structures.

Taking an anatomical view at a typical leaf like that of *F. sylvatica,* the interior of the leaf may be described as a spongy tissue (the photosynthetically active mesophyll) consisting of relatively large cells surrounded by tortuous air-filled pores (the intercellular air space). Water-saturated cell walls envelop the individual cells. The cell walls of adjacent cells are, at certain points, in contact with each other and thus form a continuous extracellular aqueous phase consisting of interconnected cell walls (called the apoplast). The individual cells are delimited by a biomembrane (plasmalemma) made up of acyl glycerol lipids and proteins. The interior of a fully differentiated leaf cell consists mainly of water contained in the very large central vacuole and in the cytoplasm. Vacuoles and organelles are surrounded by lipid membranes.

The mesophyll of leaves is bounded by a continuous cell layer, the epidermis, which separates the interior of the leaf from the atmosphere. The outer epidermal cell walls are covered by the cuticle (cutin and cuticular waxes). Adjustable pores in the epidermis (stomata) connect the intercellular air space with the surrounding atmosphere.

From this extremely simplified outline of the overall anatomy of a leaf it becomes clear that the individual compartments are in no way equally accessible to molecules coming from the outside. The cuticle is the only compartment directly exposed to the environment in its entirety and under any circumstances. This is a necessary consequence of the natural function of the plant cuticle, which is to limit the loss of water and solutes from the interior of the leaf to the environment.[31,32] Organic molecules arriving at the leaf surface as solutes or as particulates have to penetrate the cuticle in order to reach the mesophyll of the leaf. Organic vapors may take two parallel pathways; they either also penetrate the cuticle or diffuse in the vapor phase across the stomatal pores into the continuous intercellular air space. The contribution of the latter pathway depends on the degree of stomatal opening and thus is related to the physiological status of the leaf.[23]

When the molecules have reached the cell walls of the interior tissues (outer epidermal cell wall or any cell wall in contact with the intercellular air space), either they can proceed within the aqueous apoplast or they can penetrate it and then move into and across the biomembranes lining further aqueous compartments.

These considerations lead to the conclusion that an understanding of the exchange processes between the atmospheric environment and leaves and its successful modeling will be achievable only if the transport properties of the different pathways into the leaf are known. When discussing these processes special emphasis will have to be put on the roles of cuticles and stomata.

B. TRANSPORT PROPERTIES OF THE LEAF/ATMOSPHERIC ENVIRONMENT INTERFACE

1. *Fundamentals of Nonelectrolyte Transport across Plant Cuticles*

The transport of organic nonelectrolytes across the plant cuticle can be treated in a manner analogous to that across any interface separating two reservoirs. For the sake of simplicity, let us first assume that the physical states of both reservoirs are equal, i.e., that uptake occurs from an outer aqueous phase across the cuticle into the aqueous apoplast of the leaf interior. Then, the flow N (in mol · s^{-1}) of a solute from the outer reservoir (*out*) across an exposed area A of the cuticle into the inner reservoir (*in*) is given by

$$N = P_C A \left(C_W^{out} - C_W^{in} \right) \qquad (19)$$

where P_C is the permeance of the cuticle (in m · s^{-1}).[33,34] The driving force is expressed by the difference of the concentrations in both reservoirs. For the simple steady-state case, where $C_W^{out} \gg C_W^{in}$ and C_W^{out} = const., Equation 19 reduces to

$$N = P_C A C_W^{out} \qquad (20)$$

The permeance is a composite entity suitable for the phenomenological description of cuticular penetration. It takes the cuticle on an as is basis and does not normalize to unit membrane thickness. This is the pragmatic advantage of P_C since the thickness of the cuticle, especially that of the actual transport barrier within it (see Section III.B.3), is difficult to estimate. Thus, permeances are analogous to mass transfer coefficients.[35]

Despite this fact, a separation of the different components contributing to P_C may help in understanding the individual factors determining the transport process. When, for instructive purposes, the cuticle is assumed to be a homogeneous membrane, the permeance term in Equations 19 and 20 can be expanded to

$$P_C = \frac{D_C K_{CW}}{\Delta x_C} \qquad (21)$$

with D_C representing the diffusion coefficient (in m^2 · s^{-1}) of the permeating molecule in the cuticle and Δx_C the thickness (more accurately, the actual path length of diffusion) of the cuticle. This relationship shows that P_C and, according to Equation 20, also the flux across the cuticle are proportional to the mobility (D_C) and the relative solubility (K_{CW}) of the permeant in the cuticle. Thus, differences in the permeability of different chemicals across the cuticle of a given species (Δx_C is constant) may be due to different values of D_C and/ or of K_{CW}.

Together with the appropriate driving force, the permeance P_C discussed so far describes the flux of a solute across the cuticle separating two aqueous phases. Thus, P_C would be useful in describing the uptake of organic compounds arriving at the leaf surface dissolved in rain, fog, dew, or pesticide spray droplets. Under temperate climatic conditions, however, leaf surfaces are dry for considerable periods of time and then transport occurs from the vapor phase of the atmosphere across the cuticle into the aqueous phase of the apoplast.

When the outside reservoir is the atmosphere, cuticular penetration is accompanied by a phase transfer from the gaseous state to the liquid phase of the apoplast. Now the problem arises how to define correctly the driving force in Equations 19 and 20 since, obviously, concentrations in the aqueous and the vapor phase are not simply equivalent. This difficulty is resolved by transforming the vapor-phase concentration (C_A) to the equivalent aqueous concentration by $C_W = C_A/K_{AW}$ (Equation 2). Now Equation 20 can be rewritten to

$$N = P_C A \frac{C_A^{out}}{K_{AW}}$$

$$= \frac{P_C}{K_{AW}} A C_A^{out}$$

$$= g_C A C_A^{out} \tag{22}$$

where g_C now stands for conductance (mass transfer coefficient based on concentrations in the vapor phase) of the cuticle. Here, the symbol g and the term *conductance* are employed in accordance with the extended literature treating leaf gas exchange.[23]

The general procedure for converting aqueous-based permeances to mass transfer coefficients based on some driving force becomes clear when the relationship $g_C = P_C/K_{AW}$ is substituted into Equation 21:

$$g_C = \frac{D_C K_{CW}}{\Delta x_C K_{AW}}$$

$$= \frac{D_C K_{CA}}{\Delta x_C} \tag{23}$$

where K_{CA} is obtained according to Equation 5. This shows that dividing the aqueous-based permeance by K_{AW} adjusts the relative solubility parameter K_{ij} to the correct phases, namely cuticle and air instead of cuticle and water. This formalism follows Equation 4 and thus can be easily used for any desired transformation.

2. The Permeability of Plant Cuticles to Organic Compounds

The flow of organic nonelectrolytes across isolated plant cuticles can be measured experimentally and P_C derived according to Equation 20.[13,36] This has been done for a number of chemicals of diverse structures and physico-chemical properties.[10,34,37-39] The values of P_C obtained so far range over four and a half orders of magnitudes, from 1.43×10^{-11} for Primicarb to 8.60×10^{-7} for HCB (Table 8). A comparison with the respective cuticle/water partition coefficients (Table 2) shows that P_C values are heavily influenced by the K_{CW} term in Equation 21.

There is also considerable variation among P_C values determined for one chemical and the cuticles of a number of plant species. A study with 2,4-D and the cuticles from 11 plant species showed that permeances ranged over two orders of magnitude.[34] The cuticles of *Citrus aurantium* were at the lower limit of the range, while those of fruits of *Capsicum annuum* formed the upper limit. Therefore, in the following sections, permeances will be assumed to vary by a factor of 100 among species and, when appropriate, estimates will be given for the extreme values.

The observation that P_C are correlated to K_{CW} again led to the attempt to establish quantitative relationships for predicting permeances of *Citrus aurantium* leaf cuticles from more fundamental and easily accessible parameters. Such correlations have been successfully established between P_C and K_{OW}

$$\log P_C = 0.704 \log K_{OW} - 11.2 \ (r = 0.91) \tag{24}$$

and between P_C and K_{CW}

$$\log P_C = 0.734 \log K_{CW} - 11.3 \ (r = 0.95) \tag{25}$$

Using K_{CW} instead of K_{OW} somewhat increases the predictive power of the relationship.[13,38] It should be stressed in this context that Equations 24 to 26 have been derived for *Citrus aurantium* leaf cuticles and may give erroneous results for different types of cuticular membranes. Again, attention should be paid to the fact that permeances may vary between different plant species by up to two orders of magnitude.[34]

However, as predicted by Equation 21, lipophilicity cannot account for the total variation observed in the permeances of cuticles. Different molecular sizes and geometries should affect the mobility in the cuticular barrier and,

Table 8. Permeances (P_C) of *Citrus aurantium* leaf cuticles[a]

Compound	P_C Measured (m · s^{-1})	Estimated[b] (m · s^{-1})
Primicarb	—	1.43×10^{-11}
4-NP	1.62×10^{-10}	—
Atrazine	1.01×10^{-10}	—
1-NAA	3.90×10^{-9}	—
2,4-D	2.80×10^{-10}	1.93×10^{-9}
Bentazone	—	2.32×10^{-9}
Birlane	—	2.65×10^{-10}
2,4,5-T	5.20×10^{-10}	—
Triadimenol	—	7.27×10^{-10}
Tebuconazole	—	1.14×10^{-9}
WL 110547	—	3.46×10^{-9}
Bitertanol	—	6.19×10^{-10}
PCP	3.98×10^{-8}	2.09×10^{-7}
HCB	8.60×10^{-7}	—
Perylene	1.61×10^{-7}	—
DEHP	3.24×10^{-7}	—

[a] Data for 25°C compiled from References 10, 33, 37, and 38; where appropriate, values of P_C are given for the nondissociated species of the compound. See Table 2 for compound formulas.

[b] Permeances were estimated from the rate constants of unilateral desorption from the outer side (k), K_{CW}, and the thickness of the cuticle ($x_C = 2.5 \times 10^{-6}$ m) according to $P = k \times \Delta x_C \times K_{CW}$.[10]

thus, the overall transport properties of the cuticle. A comparison of Table 8 with Table 2 shows that permeances do not follow the ranking of the chemicals according to the cuticle/water partition coefficients. Quantitative evidence for the interference of the mobility term in P_C can be obtained from another property-property relationship (again for *Citrus aurantium* leaf cuticle), introducing the molar volume (LeBas) of the molecules, V_M, as an attempt to incorporate a size-related parameter:[13,38]

$$\log P_C = 238 \frac{\log K_{CW}}{V_m} - 12.5 (r = 0.98) \qquad (26)$$

Including V_m further increases the degree of determination as compared to Equation 25.

Recently, additional experimental evidence for the contribution of solute mobility to cuticular permeability has been obtained.[10] Cuticles were loaded with organic compounds and subsequently desorbed exclusively from the outer surface. The rate constants (k) determined for this unilateral desorption process are mobility parameters directly related to the diffusion coefficient in the cuticular transport barrier and have the advantage of not being affected by the relative solubility (K_{CW}) of the compound in the cuticle. The rate constants for

Citrus aurantium leaf cuticles strongly depended on the molar volume (McGowan's characteristic volume V_x) according to

$$\log k = 3.25 - 4.52 \log V_x \ (r = -0.863) \tag{27}$$

Thus, mobility decreased by a factor of 264 when molar volumes increased by only a factor of three. This demonstrates that correlating kinetic parameters for the atmosphere/leaf transfer of organic chemicals only to relative solubilities (e.g., 1-octanol/air partition coefficients[40]) may unduly oversimplify the factors affecting this process and consequently may lead to potentially erroneous conclusions.

3. The Nature of the Cuticular Transport Barrier

It has been pointed out that cuticular transport can only be described phenomenologically using permeances or conductances instead of any more formal treatment involving diffusion coefficients and path lengths because cuticles are extremely heterogeneous membranes. In the direction of atmosphere-to-leaf transport all cuticles studied so far were found to be heterogeneous with regard to fine structure, chemical composition, and transport properties.[34,41,42] Conspicuous differences in the transport properties of the outer and inner parts of a plant cuticle became evident from the following experiment: isolated cuticular membranes were preloaded with ^{14}C-labeled 2,4-D and subsequently mounted between two aqueous compartments. The desorption of 2,4-D with a buffer solution proceeded at drastically different rates to the physiological outer and inner sides.[41] The efflux rates across the inner surface were 50 to 80 times larger than those across the outer surfaces. During a 6-h desorption period only 2 to 5% of the total amount of 2,4-D initially contained in the cuticle was desorbed toward the outer side. In contrast, 75% of the amount had left the cuticle via the inner side.

This observation led to a view of the plant cuticle[32,43] which, in some way, takes up the theme of a hierarchy of accessibility (Section III.A) on a microscopic scale. The transport barrier is thought to be restricted to the very outer skin of the cuticle, making up not more than 10% of its thickness. This barrier is made up of cuticular waxes embedded within the cutin matrix of the cuticle, forming crystalline layers of low permeability.[32,44,45] After extraction of cuticular waxes by organic solvents the permeance of the cuticle to organic solutes may increase by up to three orders of magnitude (4-NP and *Citrus aurantium* leaf cuticle[38]).

The remaining inner portion of the cuticle, termed the inner volume element,[43] mainly consists of the cutin matrix. Its permeability is much higher than that of the outer skin. Even though for this reason it will not act as a transport barrier, it serves as a sorption compartment for lipophilic substances.

The spatial arrangement proposed consisting of a transport-limiting outer skin and a large inner capacity would ensure that organic substances, once taken up into the inner volume element, would be preferentially unloaded toward the inner leaf tissue. Kinetic asymmetry eventually could lead to an enhanced scavenging of organic compounds from the environment which would never be predictable from a pure equilibrium partitioning view.

4. The Transport Properties of Stomata

The stomatal pores are a pathway for the exchange of organic vapors parallel to that across the cuticle. Stomata are essential structures for all photosynthetically active parts of plants because they make possible the gas-phase uptake of CO_2. They occur either on both surfaces (amphistomatous leaf) or on only one surface of the leaf, in most cases being the lower one (hypostomatous leaf). The degree of opening of the stomatal pores depends on environmental conditions and the physiological status of the underlying tissue. Stomata are partially or completely closed during periods of water stress and, with most plant species, at darkness.[23]

For the present topic of the transfer of organic substances across the leaf/atmosphere interface only a few basic concepts out of the vast field of stomatal transport and regulation need to be considered. Basically, vapor-phase transport through the stomata is the diffusion of the molecules in an elliptical pore of depth $x_S \approx 20 \times 10^{-6}$ m (Table 4) and short and long axis on the order of 5×10^{-6} to 15×10^{-6} and 20×10^{-6} m, respectively.[23] Following Reference 23, let us call $a_S = A_S/A_L$ the fractional area of one open stoma and n the number of stomata of the leaf. Thus, na_S gives the portion of the total leaf surface area made up by stomatal pores when all stomata are open. Depending on plant species na_S ranges from 0.002 to 0.02.

Thus, the stomatal conductance of the leaf, g_S, can be defined as[23]

$$g_S = \frac{D_A na_S \alpha}{x_S + y_S} \tag{28}$$

where D_A is the diffusion coefficient of the compound in air and α is the mean degree of opening of the stomatal pores ($0 \leq \alpha \leq 1$). The effective path length experienced by the diffusing molecule is given by $x_S + y_S$, where the effective pore radius y_S corrects for the concentration patterns extending on both ends of the pore (a representative value of y_S is 5×10^{-6} m; Table 4). When D_A is not known it can be roughly estimated from the air diffusion coefficient of a substance (e.g., H_2O) according to[46]

$$D_A = D_A^{H_2O} \sqrt{\frac{MW^{H_2O}}{MW}} \tag{29}$$

Assuming that all stomata are open ($\alpha = 1$), a stomatal conductance on the order of 7×10^{-4} m · s^{-1} is obtained for a compound with a molecular weight (*MW*) of 300 g · mol^{-1}. In comparison, cuticular conductances range from 8×10^{-3} to 2×10^{-7} m · s^{-1} for *Citrus aurantium* (Table 9) and thus may go up to 0.8 m · s^{-1} for the most permeable cuticles, like those of *Capsicum annuum*. Stomatal conductances, therefore, are within the range of cuticular conductances, and the relative importance of both pathways will depend on the properties of the chemical and of the leaf/atmosphere interface (see discussion in the following section).

5. Integrating Cuticular and Stomatal Pathways of Vapor-Phase Uptake

The next step in analyzing the exchange of organic vapor across the leaf/atmosphere interface is to integrate the properties of the cuticular (Sections III.B.1 to III.B.3) and stomatal pathways (Section III.B.4) into a model for the whole process (Figure 3). Both pathways are in parallel such that individual conductances add up to the conductance of the leaf surface, g_L, according to

$$g_L = g_C + g_S \qquad (30)$$

The conductance g_L, however, does not yet fully describe the transport across the interface into the turbulent atmosphere. Unstirred layers of air covering the surface of the leaf represent a further resistance diffusing molecules experience before they enter the leaf. The thickness of this boundary layer, x_B, varies with wind speed, leaf size, and roughness of the leaf surface.[23] A representative value of x_B for leaves like those of *F. sylvatica* and a wind speed of 0.8 m · s^{-1} is approximately 1 mm. The conductance of the boundary layer, g_B, thus can be obtained from the equation

$$g_B = \frac{D_A}{x_B} \qquad (31)$$

Typical values for g_B are in the range of 6×10^{-3} m · s^{-1} and consequently are about one order of magnitude higher than those estimated previously for the stomatal pathway. A comparison with cuticular conductances for selected chemicals (Table 9) shows that the highest of them (4-NP, PCP) are of the same order of magnitude as g_B.

Boundary layer conductances may, therefore, contribute significantly to the transport properties of the atmosphere/leaf interface. The conductance of

Table 9. Cuticular permeances (P_c), cuticular (g_c) as well as total interfacial conductances (g_T), and half-times of equilibration ($t_{0.5}$) of the reference compounds[a]

Compound	PC ($m \cdot s^{-1}$)	gC ($m \cdot s^{-1}$)	gT ($m \cdot s^{-1}$)	$t_{0.5}$ (h) Citrus	$t_{0.5}$ (h) Capsicum
Phenol	4.77×10^{-10}	1.25×10^{-5}	7.56×10^{-4}	0.514	0.239
2-NP	1.02×10^{-10}	1.99×10^{-7}	6.13×10^{-4}	0.055	0.064
4-NP	1.63×10^{-10}	7.98×10^{-3}	4.41×10^{-3}	203	101
2,4-D	2.80×10^{-10}	1.25×10^{-6}	4.88×10^{-4}	0.340	0.323
Atrazine	1.01×10^{-10}	8.80×10^{-4}	1.18×10^{-3}	261	46.0
2,4,5-T	5.20×10^{-10}	2.22×10^{-4}	6.39×10^{-4}	111	13.8
PCP	3.98×10^{-8}	2.24×10^{-3}	1.91×10^{-3}	46.3	14.0
HCB	8.60×10^{-7}	2.98×10^{-4}	6.76×10^{-4}	16.7	2.17
Perylene	1.61×10^{-7}	9.05×10^{-4}	1.15×10^{-3}	1170	218
DEHP	3.24×10^{-7}	4.27×10^{-5}	4.03×10^{-4}	1140	186

[a] All permeability parameters given are for *Citrus* leaf cuticle; P_C from References 34 and 38; $t_{0.5}$ for *Capsicum* fruit cuticles were calculated assuming that their permeability was 100 times higher than that of *Citrus* leaf cuticles. See Table 2 for compound formulas.

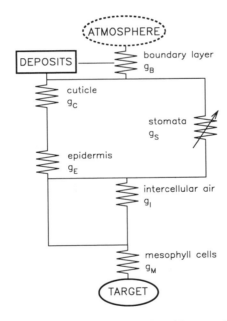

FIGURE 3. The network of resistances encountered by organic molecules diffusing from the environment (atmosphere; solid or liquid deposits on the leaf surface) to any target site within the mesophyll. The transport properties of the different barriers are characterized by the appropriate conductances g_i. The conductance of the stomata is actively controlled by the plant depending on physiological and environmental factors. Boundary layer conductance varies with wind speed and leaf surface topography.

the whole interphase, g_T, is the sum of g_L and g_B. Being conductances in series they combine according to

$$g_T = \frac{1}{\dfrac{1}{g_B} + \dfrac{1}{g_L}}$$

$$(32)$$

Total conductances for the atmosphere-to-leaf transfer of the reference chemicals now can be estimated from Equations 28 to 32 using experimentally determined permeances P_C (converted to g_C by Equation 23), the leaf properties as given in Table 4, and a D_A^{H2O} of 2.5×10^{-5} m² · s⁻¹ (Table 9). A further assumption is that the stomata of the leaf are, on the average, open during half of the period of exposure.

Now the relative importance of the cuticular and stomatal pathway can be easily assessed. The cuticular component of the total flow of a chemical in the vapor phase from the turbulent atmosphere into a leaf similar to that of *F. sylvatica*, N_C/N_T, is given by the ratio $g_{C,B}/g_T$, where $g_{C,B}$ is the total conductance of the cuticular pathway ($1/g_{C,B} = 1/g_C + 1/g_B$). For the reference chemicals and a cuticular conductance at the lower limit of the range (*Citrus aurantium*), these ratios vary from practically 0 to almost 1 (Figure 4). The cuticular pathway is predicted to be predominant during the vapor-phase uptake of 4-NP, atrazine, PCP, and perylene. When the cuticle is 100 times more permeable (*Capsicum annuum*) for all but three of the reference chemicals, $N_C/N_T \geq 0.90$. This suggests that with most lipophilic chemicals and most plant species at least half of the total flow from the atmosphere into the leaf will enter via the cuticle.

At first sight, the variation of N_C/N_T (for low cuticular permeability) may appear somewhat erratic, making desirable a more systematic exploration of the dependence of this ratio on the properties of the chemicals. Assuming that the permeability of *F. sylvatica* leaf cuticles is comparable to that of *Citrus aurantium* leaves and that Equation 25 gives realistic approximations of P_C over a wide range of K_{OW}, the variation of N_C/N_T with K_{OW} and K_{AW} can be estimated (Figure 5A). It is predicted that volatile compounds with log $K_{AW} \geq$ −1 will enter the leaf exclusively via the stomata. Within the range of realistic values of K_{OW}, lipophilicity will have no effect. When volatility decreases, the cuticular pathway will become increasingly important. More than 50% of the total flux of compounds with values of log K_{AW} of −3, −5, and −7 will enter the leaf via the cuticle when their log K_{OW} values are approximately 6.7, 4, and 1, respectively (Figure 5A).

Assuming a cuticular permeability 100 times higher than that of *Citrus aurantium,* a similar pattern, albeit with a drastically increased cuticular contribution to the total flux, is obtained (Figure 5B). Under such conditions, scarcely volatile compounds (log $K_{AW} \leq$ −5) will enter the leaf exclusively via

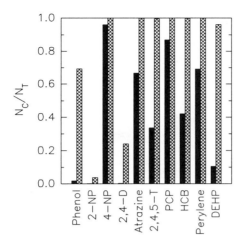

FIGURE 4. The relative importance of the flows across the cuticle and the stomata during the vapor-phase uptake of organic compounds into leaves. The ratios of cuticular flow to total flow (N_C/N_T) were calculated for leaves with low-permeability cuticles like those of *Citrus aurantium* (solid bars) and with cuticular permeabilities 100 times higher (hatched bars). See Table 2 for definitions of chemical abbreviations.

the cuticle when their log $K_{OW} \geq 4$. At least half of the total flux of compounds with log $K_{OW} \geq 1$ will already be contributed by the cuticular pathway.

This clearly emphasizes the importance of cuticular penetration for the uptake of the large number of organic pollutants and active ingredients of pesticidal formulations that have low vapor pressures. In many cases, the physicochemical properties of the compounds and the transport properties of leaf surfaces keep open only one port of entry into plants, namely the diffusion across the cuticle.

C. THE MOVEMENT OF ORGANIC COMPOUNDS WITHIN INTERIOR LEAF TISSUES

As outlined in Section III.A leaves are highly organized structures imposing a hierarchical order on the accessibility to chemicals originating from the surrounding atmosphere. Up to now this chapter has dealt with the transport properties of the immediate atmospheric environment/leaf interface. The question arises, however, whether this is indeed the rate-limiting step during the uptake of organic compounds into leaves or whether there are additional interior barriers to be considered. It might be conceivable that the further diffusion of some chemicals through the comparably thick mesophyll tissue underneath proceeds more slowly than the uptake across the interface and thus may control the rate of the whole process. A semiquantitative analysis will help to resolve this question.

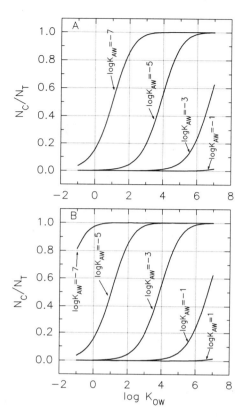

FIGURE 5. The relative importance of the flows across the cuticle and the stomata during the vapor-phase uptake of organic compounds into leaves as functions of the 1-octanol/water (K_{OW}) and air/water (K_{AW}) partition coefficients. The ratios of cuticular flow to total flow (N_C/N_T) were calculated for leaves (A) with low-permeability cuticles like those of *Citrus aurantium* and (B) with cuticular permeabilities 100 times higher.

First let us consider molecules entering the leaf via the stomatal pores and diffusing into the mesophyll along the network of the intercellular air space (Figure 3). For estimating the conductance of the intercellular air space, g_I, it is assumed that in hypostomatous leaves the fraction of the cross-sectional area of the leaf available for the gas-phase transport of molecules entering through the stomatal pores is $a_I \approx v_A/2 = 0.15$. Assuming further, conservatively, that the molecules have to cross the entire thickness of the leaf ($x_L = 2 \times 10^{-4}$ m) and that, due to tortuosity, the path length is 50% larger, an effective path length for diffusion in the intercellular air space of 3×10^{-4} m is obtained. In analogy to Equation 28, $g_I = 3 \times 10^{-3}$ m · s^{-1}. This conservative estimate is fivefold larger than the conductance for the complete stomatal pathway, $1/g_{B,S}$ $= 1/g_B + 1/g_S$ (6.3×10^{-4} m · s^{-1}), which leads to the conclusion that diffusion in the intercellular air space will not limit vapor-phase uptake through the stomata.

Now we turn to the cuticular pathway where the molecules, having just crossed the inner surface of the cuticle, enter the epidermis and the mesophyll (Figure 3). Diffusion coefficients for solutes in the aqueous phase of the cytoplasm are in the range of 10^{-9} m$^2 \cdot$ s^{-1}, and those in the cell walls are about 10^{-12} m$^2 \cdot$ s^{-1}.[47] Organic compounds in biomembranes (thickness about 5×10^{-8} m) have diffusion coefficients on the order of 10^{-14} m$^2 \cdot$ s^{-1}.[48] The conductance of the very thin biomembranes is fairly high for lipophilic organic molecules since, in analogy to Equation 21, the large partition coefficient will compensate for relatively low diffusion coefficients. Thus, in a first approximation, the resistance provided by biomembranes can be disregarded for lipophilic compounds. It remains the aqueous path which, over most of its length, crosses the aqueous phases of cytoplasm and vacuoles. In comparison to the cell lumen the thickness of the cell wall is small. Therefore, it seems reasonable to assume that the diffusion coefficient for the pathway is close to that in pure water. In the following a diffusion coefficient for the mesophyll of $D_M = 5 \times 10^{-10}$ m$^2 \cdot$ s^{-1} will be used.

Now the permeance of the continuous cell layer directly adjacent to the cuticle, the epidermis, can be estimated. The path length for diffusion across it is about 2×10^{-5} m; thus, a permeance of 2.5×10^{-5} m $\cdot$ s^{-1} is obtained. Comparing this permeance to the P_C of a low-permeability cuticle (which again is assumed to depend on K_{OW} as given by Equation 24), it is evident that, even with very lipophilic compounds, cuticular permeance is at least one and a half orders of magnitude lower (Figure 6). This means that penetration through the epidermis will not be rate limiting during uptake. Even when cuticular permeability is 100 times higher than that found with *Citrus aurantium,* epidermal permeance will be lower than P_C for all compounds with log $K_{OW} \leq 6.5$ (Figure 6).

Let us now extend this consideration to the case where, hypothetically, the sink for the material entering the leaf through the cuticle is located on a plane exactly in the middle between the two leaf surfaces. The path length then would be approximately 1×10^{-4} m, and a fraction of $a_M = 0.7$ of the cross-sectional area would be available for the diffusion in the aqueous phase of the mesophyll. Analogous to Equation 28, a permeance of the mesophyll of 3.5×10^{-6} m $\cdot$ s^{-1} is now obtained. This permeance is at least a factor of ten above the P_C of low-permeability cuticles for all solutes with log $K_{OW} \leq 6.5$ (Figure 6). The permeance of high-permeability cuticles like those of *Capsicum annuum* will be smaller than the permeance of the mesophyll for compounds with log $K_{OW} \leq 7$ (Figure 6).

The conclusions that can be drawn from this comparison of interfacial transport and internal transport are important: (1) the crossing of the atmospheric environment/leaf interface either by the stomatal or the cuticular pathway will be the rate-limiting step for the vast majority of compounds and plant species, and (2) any solutes having left the inner surface of the cuticle will

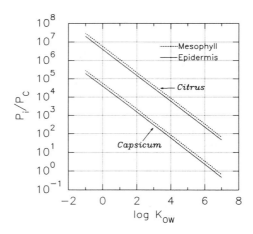

FIGURE 6. Permeances of interior tissues of leaves in relation to the permeances of cuticles like those of *Citrus aurantium* and those of *Capsicum annuum* (P_c 100 times higher than that of *C. aurantium*). The ratios of epidermal or mesophyll permeance to cuticular permeance (P_i/P_c) are shown as functions of the 1-octanol/water partition coefficients of (K_{ow}) organic chemicals.

move across the mesophyll almost instantaneously in the time scale of cuticular transport. The consequence of the latter conclusion is that the inner volume element of the cuticle (see Section III.B.3) will be at partitioning equilibrium with the mesophyll. The unloading of molecules penetrating the cuticle is not kinetically hindered and, thus, no local accumulation in the inner volume element of the cuticle in excess of the overall partitioning equilibrium within the leaf will occur.

The comparably rapid equilibration within the mesophyll also has an important consequence for those cases where the compound under consideration is not persistent but is subject to conjugation, metabolism, translocation, or growth dilution within the interior leaf tissues. A rate of removal equal to or higher than that of cuticular entry would keep concentrations in the mesophyll and, consequently, in the inner volume element of the cuticle at a very low level down to practically zero. The same would be the case with weak electrolytes which are completely ionized in one of the aqueous compartments of the leaf tissue. Whenever a mechanism of removal is active within the leaf, the concentration of a compound in the tissue or the cuticle will not depend in a direct and simple way on its concentration in the atmospheric environment. Thus, leaves, like any other living parts of plants, are in principle unsuitable means for biomonitoring environmental background concentrations of most organic compounds.

D. TIME COURSES OF VAPOR-PHASE UPTAKE INTO LEAVES

After learning about the transport properties of the atmospheric environment/leaf interface and evaluating the further diffusion within the interior of the leaf, it is time to focus on more general aspects of the vapor-phase uptake of organic compounds into leaves. In the introductory remarks to Section III it has been mentioned that the equilibrium view of atmospheric environment/leaf exchange assumes that equilibration times are negligible within the appropriate time scale. Using the information presented in the preceding sections it is now possible to assess quantitatively whether this assumption is realistic.

For this purpose it is assumed that at time $t = 0$ the concentration of a chemical in the leaf, C_L^0, is zero and that this leaf is suddenly brought into contact with air of concentration C_A, the latter being constant throughout the experiment. The time course of the uptake of the chemical into the leaf is given by[16]

$$\frac{C_L^t}{C_L^\infty} = 1 - \exp\left(-\frac{g_T A_L t}{V_L K_{LA}}\right)$$

$$= 1 - e^{-kt} \tag{33}$$

where g_T is the total interfacial conductance according to Equation 32, A_L and V_L are the total leaf area and volume, respectively, and k is the rate constant of the process. The term on the left side of Equation 33 is the ratio of the concentration in the leaf at any given time t and the concentration in the leaf at equilibrium ($C_L^\infty = K_{LA} C_A$).

Using Equation 33 and assuming that conditions are permanently favorable for vapor-phase uptake, it is now possible to calculate the time course of equilibration of a chemical between the atmosphere and a leaf similar to those of *F. sylvatica*. As an extreme example it is evident that under these conditions it will take about 500 days before a leaf having a cuticle with properties like that of *Citrus aurantium* will be in equilibrium with the atmospheric concentration of perylene (Figure 7). A more instructive and reliable measure of the velocity of equilibration is the half-time ($t_{0.5}$) of this process, given by

$$t_{0.5} = -\frac{\ln 0.5}{k} \tag{34}$$

Assuming a low-permeability cuticle (Table 9), half-times range from 3.3 min (2-NP) to 49 days (perylene); they range from 3.8 min to 9 days for the same compounds if the cuticular permeance is 100 times higher than that of *Citrus aurantium*. As a rule of thumb, equilibration will be nearly completed after ten half-times. Thus, one vegetation period of 180 days (typical for latitudes at approximately 40°) will be sufficient for the equilibration of most

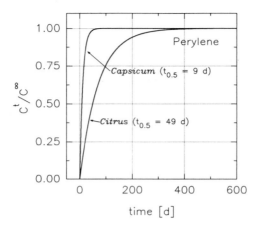

FIGURE 7. Time courses for the uptake of vapor-phase perylene into leaves. The ratios of the concentration in the leaf at a given time over the equilibrium concentration (C^t/C^∞) are plotted for low-permeability (*Citrus*) and high-permeability (*Capsicum*) cuticles, respectively.

of the chemicals and the leaves of plant species with high cuticular permeabilities. With low P_C, one vegetation period would not suffice for vapor-phase perylene and DEHP to achieve equilibrium between the atmosphere and leaf. In this case, concentrations of atrazine and 4-NP in the air should remain constant for a period of time at least half as long as the vegetation period. Obviously, with some compounds, equilibration times will be longer than the usual time scale of the variation of the vapor-phase concentration of organic compounds in the environment.

The dependence of the half-times of vapor-phase uptake into *F. sylvatica* leaves on the physicochemical properties of organic compounds also can be examined systematically when the usual assumption concerning the relationship between K_{OW} and P_C of *Citrus aurantium* cuticles is made. Below log K_{OW} ≈ 3 the half-times are only affected by the volatility of the compounds (Figure 8). When K_{AW} decreases from 1 to 10^{-7}, the half-times for chemicals with K_{OW} ≤ 3 will increase from several seconds to more than a year. The half-times of more lipophilic compounds (log K_{OW} ≥ 3) increase with increasing K_{OW} and decreasing K_{AW}. The same pattern of dependence of the half-times on the physicochemical properties of the compounds is found when a cuticular permeance 100 times higher than that of *Citrus aurantium* is assumed. In this case, the family of curves, as shown in Figure 8, is just moved to lower half-times.

This characterization of the kinetics of uptake into leaves clearly shows that the leaves of many plant species will attain equilibrium with the atmospheric concentrations of a large number of organic chemicals of intermediate lipophilicity and volatility. There may be cases due either to low cuticular permeability or to extremely low volatility and high lipophilicity, however,

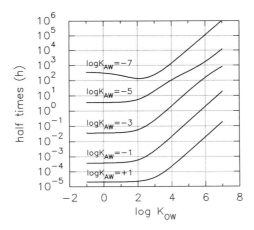

FIGURE 8.   The half-times of vapor-phase equilibration of organic chemicals between the atmosphere and a leaf as functions of the 1-octanol/water (K_{OW}) and the air/water (K_{AW}) partition coefficients. A low-permeability cuticle like that of *Citrus aurantium* is assumed.

where leaves will not reach equilibrium within one vegetation period. In these cases, interfacial transport will be too sluggish to follow the outside concentration changes. Consequently, the usefulness of leaves as devices for monitoring background contamination levels again must be questioned.

At this point still a further aspect has to be emphasized. It should be realized that everything said so far about the atmosphere-to-leaf transfer of organics equally applies to the exchange in the reverse direction. Thus, whenever atmospheric concentrations fall below $C_A = C_L'/K_{LA}$, revolatilization of the compound will set in. This makes the interpretation of the leaf/atmospheric environment system even more complex. From analyzing the leaf concentration of a compound having a long half-time one can never decide whether the leaf is indeed at equilibrium or whether the data come just from an arbitrary point of an uptake or revolatilization time course like that in Figure 7. Thus, in conclusion, data on the concentration of environmental organic compounds in leaves can provide useful information only when (1) the rates of metabolism, conjugation, translocation, and growth dilution; (2) the history of concentration changes in the atmospheric environment; and (3) the extent of the preloading of the leaf are known. How far this knowledge has to extend into the past is dictated by the equilibration kinetics of the specific compound and the leaf under study.

E. UPTAKE FROM AQUEOUS AND PARTICULATE DEPOSITS ON LEAF SURFACES

The emphasis of the preceding treatment of the time courses of uptake of organic compounds from the atmospheric environment into leaves (Section

III.D) has been placed on chemicals present in the vapor phase. The reason for this is only partially the undisputable importance of this mechanism of atmosphere-to-vegetation exchange of organic chemicals. Equally important for this decision to concentrate on the vapor phase is the relative simplicity of the processes involved. Most of the general aspects presented so far, however, apply both to the vapor-phase uptake and to the transfer of organic chemicals originating from aqueous solutions or particulate deposits on the leaf surfaces (Figure 3). This holds for the consideration about the accessibility of the different leaf compartments (Section III.A), the outline of cuticular transport properties (Section III.B), and the discussion of the movement of organic molecules in the aqueous phase of the mesophyll (Section III.C).

High concentrations of lipophilic organic compounds occur in rain and fog water. Urban fog has been shown to be an especially efficient scavenger for lipophilic chemicals in the atmosphere,[49,50] and concentrations in fog water ranging from 1.7 to 7.5 ng · m^{-3} have been determined for total polychlorinated biphenyls (PCBs) and *n*-alkanes.[50] Varying amounts of the highly lipophilic material in fog are associated with the particles suspended within the liquid phase. The fractions associated with the particle phase correlated inversely with water solubilities and ranged from 10 to 70%.[50] The total deposition of organic compounds via rain and fog may range from 0.09 (chlorinated dioxins and furans) to 200 µg · m^{-2} · a^{-1} (total PCBs).[50] Deposition via fog was found to be 100 times more important than that via rain.

Particulate aerosols carrying loads of adsorbed organic compounds are also deposited onto leaf surfaces. The importance of particle-bound input into vegetation may be assessed from the amount of particulate matter adhering to leaf surfaces. Particle coverages of up to 5.8 × 10^{-4} kg · m^{-2} were found on 30-month-old needles of tall 90-year-old *Picea abies* (L.) Karst trees when the annual mean aerosol concentration in the air was 30 µg · m^{-3}.[51]

Qualitatively, these examples show that a considerable fraction of the total input of organic compounds into above-ground parts of vegetation will originate from liquid and dry deposition on leaf surfaces. Whenever the concentrations in the deposits exceed the equilibrium concentrations predicted by particle/air or air/water partition coefficients, a surplus flux of organic chemicals across the atmospheric environment/leaf interface will occur. Several mechanisms leading to higher-than-equilibrium concentrations in particles and small droplets in the atmosphere have been described. Therefore, the uptake predicted for vapor-phase processes alone is the minimum one.

Now let us discuss some of the additional problems that are associated with uptake from aqueous or particulate deposits and, unfortunately, forbid simple formulations of uptake kinetics like that in Equation 33. Here, the problem is that two out of three components of the general flow equation (Equation 19), namely the driving force and the area exposed, are difficult or even impossible to define under natural conditions.

Let us begin with a situation frequently found when small droplets of a solution of an organic chemical are deposited on the leaf surface. In such a case, there is no more or less homogeneous and well-defined background concentration in the practically infinite environment which could be taken as the driving force for the uptake into leaves. Instead of this, there are small limited volumes of aqueous solutions. The sessile droplets may be more or less regularly hemispherical when the cuticular surface is homogeneously smooth and hydrophobic. In most cases their volume will be small compared to the volume of the leaf. Thus, steady-state flow of the solute contained in the droplet across the cuticle into the leaf will persist only for a short period of time. Afterward, due to the limited volume of the donor, the driving force will decay exponentially with time.[33]

Another problem with uptake from the aqueous (and particulate) phase is the determination of the actual surface area exposed. One reason for this is that it is difficult to estimate the area actually wetted by precipitation, especially when the leaves are agitated and when their surfaces are contaminated by epiphyllic microorganisms. A more fundamental reason which may apply equally well to vapor-phase uptake is the microscopic roughness of the leaf surfaces of many plant species. These rough surfaces are made up of epicuticular waxes forming layers of erect platelets, rodlets, ribbons, or tubules.[52] Such layers prevent the wetting of the cuticular surface and consequently may drastically reduce the exposed area since the sessile droplets make contact only with the tips of these sculptures.

On the other hand, however, after prolonged incubation with aqueous solutions, needle surfaces covered with such structures were also observed to be wetted. As a consequence, the effective surface area, on a microscopic scale, exposed to the outside solution dramatically increases. Effective surface areas of conifer needles determined by the adsorption of PCP from aqueous solutions were 10 (*P. abies*) to 137 (*Abies koreana*) times larger than the corresponding macroscopic surface areas,[53] and it is the microscopic surface area which determines solute uptake. Rates of solute uptake into the needles were proportional to effective surface areas. Thus, when exposed to aqueous deposits of equal concentrations, leaves with large microscopic surfaces areas may accumulate appreciably higher amounts of organic compounds than those with smooth cuticular surfaces.[53]

REFERENCES

1. Schulze, E. D. (1982), Plant life forms and their carbon, water and nutrient relations, in *Physiological Plant Ecology* (Lange, O. L., Nobel, P. S., Osmond, C. B., and Ziegler, H., Eds.), Part 2, Encyclopedia of Plant Physiology, Vol. 12B, Springer-Verlag, Berlin, p. 615.

2. Larcher, W. (1984), *Ökologie der Pflanzen,* Ulmer Verlag, Stuttgart.
3. Leo, A., Hansch, C., and Elkins, D. (1971), Partition coefficients and their uses, *Chem. Rev.,* 71: 525.
4. Platford, R. F. (1979), Glyceryl trioleate-water partition coefficients for three simple organic compounds, *Bull. Environ. Contam. Toxicol.,* 21: 68.
5. Travis, C. C. and Hattemer-Frey, H. A. (1988), Uptake of organics by aerial plant parts: a call for research, *Chemosphere,* 17: 277.
6. Riederer, M. and Schönherr, J. (1984), Accumulation and transport of (2,4-dichlorophenoxy)acetic acid in plant cuticles. I. Sorption in the cuticular membrane and its components, *Ecotoxicol. Environ. Saf.,* 8: 236.
7. Kerler, F. and Schönherr, J. (1988), Accumulation of lipophilic chemicals in plant cuticles: prediction from octanol/water partition coefficients, *Arch. Environ. Contam. Toxicol.,* 17: 1.
8. Shafer, W. E. and Schönherr, J. (1985), Accumulation and transport of phenol, 2-nitrophenol, and 4-nitrophenol in plant cuticles, *Ecotoxicol. Environ. Saf.,* 10: 239.
9. Sabljic', A., Güsten, H., Schönherr, J., and Riederer, M. (1990), Modeling plant uptake of airborne organic chemicals. I. Plant cuticle/water partitioning and molecular connectivity, *Environ. Sci. Technol.,* 24: 1321.
10. Bauer, H. and Schönherr, J. (1992), Determination of mobilities of organic compounds in plant cuticles and correlation with molar volumes, *Pestic. Sci.,* 35: 1.
11. Ettlinger, K. (1992), *Das Biopolymer Kutin: Chemische Zusammensetzung und ökotoxikologische Bedeutung als Sorptionskompartiment für flüchtige organische Verbindungen,* dissertation, Technische Universität München, Munich, Germany.
12. Riederer, M. and Schönherr, J. (1986), Thermodynamic analysis of nonelectrolyte sorption in plant cuticles: The effects of concentration and temperature on sorption of 4-nitrophenol, *Planta,* 169: 69.
13. Schönherr, J. and Riederer, M. (1989), Foliar penetration and accumulation of organic chemicals in plant cuticles, *Rev. Environ. Contam. Toxicol.,* 108: 1.
14. Kier, L. B. and Hall, L. H. (1986), *Molecular Connectivity in Structure-Activity Analysis,* John Wiley & Sons, New York.
15. Bauer, H. and Schönherr, J. (1992), Unpublished results.
16. Riederer, M. (1990), Estimating partitioning and transport of organic chemicals in the foliage/atmosphere system: discussion of a fugacity-based model, *Environ. Sci. Technol.,* 24: 829.
17. Schneider, G. (1990), *Die kutikulären Wachse von* Citrus aurantium *L. und* Fagus sylvatica *L.: Einfluß des Blattalters auf Zusammensetzung und Eigenschaften,* dissertation, Technische Universität München, Munich, Germany.
18. Eschrich, W., Burchardt, R., and Essiamah, S. (1989), The induction of sun and shade leaves of the European beech (*Fagus sylvatica* L.): anatomical studies, *Trees,* 3: 1.
19. Matzke, K. and Riederer, M. (1991), A comparative study into the chemical constitution of cutins and suberins from *Picea abies* (L.) Karst., *Quercus robur* L., and *Fagus sylvatica* L., *Planta,* 185: 233.
20. Schreiber, L. and Schönherr, J. (1990), Phase transitions and thermal expansion coefficients of plant cuticles, *Planta,* 182: 186.

21. Vogt, K. A., Grier, C. C., and Vogt, D. J. (1986), Production, turnover, and nutrient dynamics of above- and belowground detritus of world forests, *Adv. Ecol. Res.,* 15: 303.

22. Sitte, P., Ziegler, H., Ehrendorfer, F., and Bresinsky, H. (1991), *Lehrbuch der Botanik für Hochschulen,* 33rd ed., Gustav Fischer Verlag, Stuttgart.

23. Nobel, P. S. (1991), *Physicochemical and Environmental Plant Physiology,* Academic Press, San Diego.

24. Rippen, G. (1991), *Handbuch der Umweltchemikalien,* ecomed-Verlag, Landsberg/ Lech.

25. Suntio, L. R., Shiu, W. Y., Mackay, D., Seiber, J. N., and Glotfelty, D. (1988), Critical review of Henry's law constants for pesticides, *Rev. Environ. Contam. Toxicol.,* 103: 1.

26. McLachlan, M. S. (1992), *Das Verhalten hydrophober chlororganischer Verbindungen in laktierenden Rindern,* dissertation, Universität Bayreuth, Germany.

27. Mackay, D. (1979), Finding fugacity feasible, *Environ. Sci. Technol.,* 13: 1218.

28. Mackay, D. and Paterson, S. (1981), Calculating fugacity, *Environ. Sci. Technol.,* 15: 1006.

29. Mackay, D. and Paterson, S. (1982), Fugacity revisited, *Environ. Sci. Technol.,* 16: 654A.

30. Mackay, D., (1991). *Multimedia Environmental Models,* Lewis Publishers, Boca Raton, FL.

31. Schönherr, J. (1982), Resistance of plant surfaces to water loss: transport properties of cutin, suberin and associated lipids, in *Physiological Plant Ecology,* (Lange, O. L., Nobel, P. S., Osmond, C. B., Ziegler, H., Eds.), Part 2, Encyclopedia of Plant Physiology, Vol. 12B, Springer-Verlag, Berlin, p. 153.

32. Riederer, M. (1991), Die Kutikula als Barriere zwischen terrestrischen Pflanzen und der Atmosphäre, *Naturwissenschaften,* 78: 201.

33. Hartley, G. S., Graham-Bryce, I. J. (1980), *Physical Principles of Pesticide Behaviour,* Academic Press, New York.

34. Riederer, M., Schönherr, J. (1985), Accumulation and transport of (2,4-dichlorophenoxy)acetic acid in plant cuticles. II. Permeability of the cuticular membrane, *Ecotoxicol. Environ. Saf.,* 9: 196.

35. Cussler, E. L. (1984), *Diffusion,* Cambridge University Press, London.

36. Kerler, F., Riederer, M., and Schönherr, J. (1984), Non-electrolyte permeability of plant cuticles: a critical evaluation of experimental methods, *Physiol. Plant.,* 62: 599.

37. Schönherr, J. (1976), Naphthaleneacetic acid permeability of *Citrus* leaf cuticle, *Biochem. Physiol. Pflanzen,* 170: 309.

38. Kerler, F. and Schönherr, J. (1988), Permeation of lipophilic chemicals across plant cuticles: prediction from partition coefficients and molecular volumes, *Arch. Environ. Contam. Toxicol.,* 17: 7.

39. Chamel, A. (1986), Foliar absorption of herbicides: study of the cuticular penetration using isolated cuticles, *Physiol. Vég,* 24: 491.

40. Paterson, S., Mackay, D., Bacci, E., and Calamari, D. (1991), Correlation of equilibrium and kinetics of leaf-air exchange of hydrophobic organic chemicals, *Environ. Sci. Technol.,* 25: 866.

41. Schönherr, J. and Riederer, M. (1988), Desorption of chemicals from plant cuticles: evidence for asymmetry, *Arch. Environ. Contam. Toxicol.*, 17: 13.

42. Riederer, M. and Schönherr, J. (1988), Development of plant cuticles: fine structure and cutin composition of *Clivia miniata* Reg. leaves, *Planta*, 174: 127.

43. Schönherr, J. Riederer, M., Schreiber, L., and Bauer, H. (1991), Foliar uptake of pesticides and its activation by adjuvants: theories and methods for optimization, in *Pesticide Chemistry* (Frehse, H., Ed.), VCH-Verlag, Weinheim, p. 237.

44. Riederer, M. and Schneider, G. (1990), The effect of the environment on the permeability and composition of *Citrus* leaf cuticles. II. Composition of soluble cuticular lipids and correlation with transport properties, *Planta*, 180: 154.

45. Reynhardt, E. C. and Riederer, M. (1991), Structure and molecular dynamics of the cuticular wax from leaves of *Citrus aurantium* L., *J. Phys. D:*, 24: 478.

46. Thomas, R. G. (1982), Volatilization from water, in *Handbook of Chemical Property Estimation Methods,* (Lyman, W. J., Reehl, W. F., Rosenblatt, D. H., Eds.), McGraw-Hill, New York, p. 15-1.

47. Canny, M. J. (1990), Rates of apoplastic diffusion in wheat leaves, *New Phytol.*, 116: 263.

48. Lieb, W. R. and Stein, W. D. (1971), The molecular basis of simple diffusion within biological membranes, in *Current Topics in Membranes and Transport* (Bronner, F., Kleinzeller, A., Eds.), Vol. 2, Academic Press, New York.

49. Leuenberger, C., Czuczwa, J., Heyerdahl, E., and Giger, W. (1988), Aliphatic and polycyclic aromatic hydrocarbons in urban rain, snow and fog, *Atmosph. Environ.*, 22: 695.

50. Capel, P. D., Leuenberger, C., and Giger, W. (1991), Hydrophobic organic chemicals in urban fog, *Atmos. Environ. A*, 25: 1335.

51. Simmleit, N., Tóoth, A., Székely, T., and Schulter, H. R. (1989). Characterization of particles adsorbed on plant surfaces, *Int. J. Environ. Anal. Chem.*, 36: 7.

52. Baker, E. A. (1982), Chemistry and morphology of plant epicuticular waxes, in *The Plant Cuticle,* Cutler, D. F., Alvin, K. L., and Price, C. E., Eds., Academic Press, London, p. 139.

53. Schreiber, L. and Schönherr, J. (1992), Uptake of organic chemicals in conifer needles: surface adsorption and permeability of cuticles, *Environ. Sci. Technol.*, 26: 153.

CHAPTER 7

Interpreting Chemical Partitioning in Soil-Plant-Air Systems with a Fugacity Model

Sally Paterson and Donald Mackay

TABLE OF CONTENTS

I. INTRODUCTION

The nature of the transport and transformation processes which occur to organic chemicals in a soil-plant-air system are of obvious agronomic relevance. Furthermore, it is increasingly recognized that plant biomass plays an important role in the cycling of organic contaminants, affecting atmospheric levels of organic chemicals both as a source and as a sink.[1] Vegetation may

1-56670-078-7/95/$0.00+$.50

serve to facilitate migration of chemicals from soil to air. Accumulation in vegetation is also an important initial step in the uptake processes of the terrestrial food web, providing a route by which birds, wildlife, agricultural animals and, eventually, humans are exposed to chemicals present in soil.

Specifically, eucalypt foliage has been used for extensive monitoring of polychlorinated biphenyls (PCBs) in Australia.[2] Bacci and coworkers have investigated the relationship between levels of chlorinated hydrocarbons in air and various plant species in both cold and tropical climates.[1,3,4] Concentrations of nitrophenols in conifers, chlorinated hydrocarbons in pine and spruce needles, and PCBs in tree bark have been related to local atmospheric levels.[5-8] Grain crops have been widely used to monitor uptake from soil or soil solution.[9-11] Mosses and lichens have been shown to be bio-indicators of the presence of organochlorines and PCBs.[12] In remote, uninhabited regions, these plants become indicators of long-range aerial transport.[13]

Chemical uptake by plants from soil and the atmosphere and subsequent distribution within various plant tissues are affected by various properties of the compound, the plant, and its environment. These factors include (1) the physicochemical properties of the substance, such as molecular weight, vapor pressure, aqueous solubility, and octanol/water partition coefficient;[10,14,15] (2) plant characteristics such as the nature of the root system, lipid or wax content, and leaf morphology; and (3) environmental characteristics such as soil organic and mineral content,[9,16] meteorology, and temperature.[17]

In view of the environmental and agricultural importance of this issue, the large number of chemicals of concern, and the wide variety of soils and plants, there is an incentive to devise general methods of quantifying or modeling these processes in the hope that they will assist in extrapolation from chemical to chemical, from plant to plant, and between environmental conditions.

Among the models which have been proposed are those of Paterson et al.,[18] which treats a plant in a manner similar to this study; Calamari et al.,[19] which predicts the equilibrium concentration of several pesticides in the biomass component of an evaluative environment; Schramm et al.,[20] which predicts the dynamic distribution in spruce needles; and Bacci et al.,[21] which describes the accumulation and release kinetics of azalea leaves exposed to constant vapor levels of organic chemicals. Trapp et al.[15] and Riederer[22] have developed fugacity-based models to describe the transport and distribution of chemicals between soil, air, water, and plant tissues.

This chapter describes and discusses a model which treats chemical migration into three plant compartments — root, stem, and foliage — from two environmental compartments, soil and atmosphere. The model includes chemical transport between air and soil and can treat chemical emissions into any or all of the five compartments. Advective inflow of chemicals in air also may be included. Rates of transport between, and transformation in, the various plant and environmental compartments are calculated. Expressions for metabolism and linear plant growth can be included but are treated only tentatively at this stage because of lack of adequate data.

The model is assembled as a set of differential equations in time which can be solved numerically. Also of interest is the steady-state solution to these equations, i.e., conditions which would prevail for long periods of time. The main purpose of this work is to suggest that the primary features of chemical behavior in a system can be assessed by inspecting and interpreting the various fugacity parameter values as well as the differential and steady-state equations and their solutions. This permits quantitative assertions to be made about the uptake and release of a variety of chemicals by different species of plants, at various stages of growth, from soil, air, and aerosol particles. It is believed that this multicompartment approach, which is used so successfully in air, water, soil, and even pharmacokinetic models, is applicable to plants.

The model described here is viewed as an evolving tool to estimate the approximate distribution and concentrations of chemical in the plant compartments as a function of time and to indicate the relative importance of soil and air as sources. Undoubtedly, as more data become available and fitted to the model, deficiencies will be identified and improvements will be made. This model of organic chemical migration in soil-plant-air systems thus represents only one stage in a continuing process.

II. FUGACITY CONCEPT

The model is formulated with the aid of the fugacity concept, which has been described extensively elsewhere.[23,24] Fugacity f, with units of pressure (Pa), is an equilibrium criterion which is linearly related to concentration C (mol $\cdot$ m^{-3}) through a fugacity capacity Z (mol $\cdot$ m$^{-3} \cdot$ Pa^{-1}), where $C = Zf$. When two phases achieve equilibrium with respect to chemical distribution, the chemical fugacities are equal and partitioning can be described in terms of their Z values, which are essentially "half" partition coefficients (K_{12}), as shown below:

$$C_1/C_2 = fZ_1/fZ_2 = Z_1/Z_2 = K_{12}$$

The fugacity capacity, or Z value, of a chemical in a phase depends on the physicochemical properties, temperature, and the nature of the phase, especially its content of air, water, organic matter, lipids, or waxes. A unique Z value generally exists for each chemical in each phase. Estimation of Z values starts in the air phase and then proceeds to water and other phases, employing correlations for the partition coefficients.

The expressions for estimating Z values for air, water, soil, and organic matter are summarized in Table 1. The detailed derivation of these expressions for plant compartments is described in Section III.B.

Table 1. Definitions and suggested correlations for Z values (mol · m^{-3} · Pa^{-1})

Compartment			
Air (Z_A)	1/RT	R	= 8.314 (Pa · m^3/(mol K))
		T	= absolute temperature K
Air particles (Z_Q)	$Z_A K_{QA}$	K_{QA}	= aerosol-air partition coefficient
		K_{QA}	$\approx 6 \times 10^6/P_L^S$
		P_L^S	= subcooled liquid vapor pressure (Pa)
Water (Z_W)	1/H or C^S/P^S	H	= Henry's law constant (Pa · m^3 · mol^{-1})
		C^S	= aqueous solubility (mol · m^{-3})
		P^S	= vapor pressure (Pa)
Soil (Z_E)	$\phi_{OC} \cdot K_{OC} \cdot \rho_E \cdot Z_W$	K_{OC}	= organic carbon partition coefficient
			$\approx 0.41 K_{OW}$
		ϕ_{OC}	= fraction organic carbon, e.g., 0.0015
		ρ_E	= soil density (kg · L^{-1})
Plant roots (Z_R)	RCF · Z_W · ρ_R/ρ_W	RCF	= 0.82 + 0.014 K_{OW}
Plant stem (Z_S)	SXCF · Z_{XY} · ρ_S/ρ_W	SXCF	= 0.82 + 0.0065 K_{OW}
Plant xylem contents (Z_{XY})	1/H or Z_W, i.e., equivalent to water		
Plant phloem contents (Z_{Ph})	1/H or Z_W, i.e., equivalent to water		
Plant leaf (Z_L)	0.18Z_A + 0.80Z_W + 0.02 K_{OW} · Z_W, i.e., air, water, octanol "mixture"		
Densities (kg/L)	$\rho_R = 0.83$	$\rho_E = 1.55$	
	$\rho_S = 0.83$	$\rho_W = 1.0$	
	$\rho_L = 0.82$		

Transport and transformation processes are expressed in terms of D values with units of mol · Pa^{-1} · h^{-1}, the rate N being Df mol · h^{-1}. These D values apply to a phase of fugacity capacity Z characterizing

1. Diffusive transfer between compartments as

$$D = kAZ \text{ (mol · Pa}^{-1} \cdot \text{h}^{-1})$$

where k is a mass transfer coefficient (MTC; m/h) and A is the interfacial area (m^2)

2. Bulk flow in a phase such as air or water as

$$D = GZ \text{ (mol · Pa}^{-1} \cdot \text{h}^{-1})$$

where G is the flow rate (m^3/h) of the phase

3. Reaction (including metabolism) as

$$D = k_R VZ \text{ (mol · Pa}^{-1} \cdot \text{h}^{-1})$$

where k_R is the first order reaction rate constant (h^{-1}), and V is the compartment volume (m^3)

Mackay[24] provides a more detailed account of the derivation and use of Z and D values.

III. MODEL DESCRIPTION

A MODEL ENVIRONMENT

The illustrative model environment, depicted in Figure 1, consists of a plant rooted in a defined volume of soil and surrounded by a defined volume of air. The soil, which is 1 m^2 in area and 5 cm in depth, consists of air, water, and organic and mineral matter. The atmosphere is considered to extend to a height of 1 m and consists of air and (if desired) aerosol particles. The plant consists of three compartments — root, stem, and foliage — with dimensions resembling those of the soybean. The roots occupy a volume of 10 cm^3, or 10^{-5} m^3. The stem has a length of 50 cm, a diameter of 0.5 cm, and a volume of approximately 10 cm^3. The foliage consists of 20 leaves each with an area of 25 cm^2 and a thickness of 0.5 mm, resulting in a total leaf volume of 25 cm^3. The assumed dimensions and properties of the various compartments are given in Table 2, but these may, of course, be varied to simulate any desired situation.

Chemical may be initially present in the soil at a defined concentration, or it may enter the environment by direct emission at a defined rate to any of the other four compartments or by advective inflow in air. The plant may take up chemical from the soil through the roots, from the air through the cuticle or stomata of the foliage, or by direct emissions onto foliage, stem, or roots. The chemical is transported through the plant by flow in the xylem and phloem. It may then partition into various plant tissues, exit through the transpiration stream, diffuse through the roots, or be metabolized. Transfer of chemical between soil and air is included which allows quantification of the rate of soil-air-foliage transfer, indicating the likely importance of this route compared to the soil-root-stem-foliage route.

B. FUGACITY CAPACITIES OR Z VALUES

The expressions used to estimate fugacity capacities (Z values) for the various plant compartments are based on the work by Briggs and coworkers[11,25] and Bacci and coworkers[14,21] and have been described previously.[18,26] These correlations are described briefly below and are summarized in Table 1.

Briggs and coworkers[11] correlated uptake of nonionic chemicals by barley with the octanol/water partition coefficient, K_{OW}, as a root concentration factor (RCF):

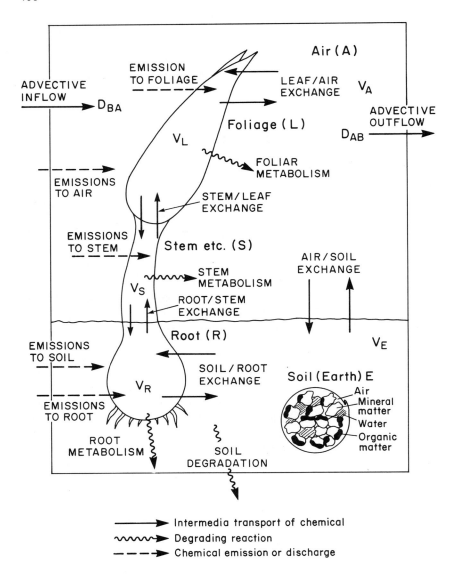

FIGURE 1. Illustrative soil-plant-air system showing compartments and transport and transformation processes.

$$\log (RCF - 0.82) = 0.77 \log K_{OW} - 1.52$$

$$RCF = C_R/C_W \text{ at equilibrium (on a mass/mass basis)}$$

C_R and C_W are chemical concentration in roots (fresh weight) and external solution, respectively. Rearranging this equation gives

Table 2. Dimensions of illustrative plant-soil-air system

Compartment	Volume (m³)	Width/diameter (m)	Density (kg · m⁻³)
Roots	10×10^{-6}	—	830
Stem	9.82×10^{-6}	0.005	830
Xylem	0.982×10^{-6}	0.00158	1000
Phloem	0.701×10^{-6}	—	1000
Leaf (20)	25×10^{-6}	5×10^{-4}	820
Air	1	$1\,m \times 1\,m \times 1\,m$	1.19
Soil	0.05	$1\,m \times 1\,m \times 0.05\,m$	1500

$$C_R = 0.82C_W + 0.03(K_{OW})^{0.77} \cdot C_W$$

The first term on the right-hand side of the equation represents the capacity of the water contained in the root for a chemical, and the second term represents the capacity of the organic root tissue. Modifying the correlation by forcing a power of 1.0 on K_{OW} results in a simple expression,

$$RCF = 0.82 + 0.014K_{OW}$$

The barley root can thus be regarded as consisting of 82% water and 1.4% octanol-equivalent organic matter. These percentages clearly vary between plant species.

Briggs et al.[25] also developed a correlation to describe partitioning between stem and xylem sap, namely SXCF:

$$\log (SXCF - 0.82) = 0.95 \log K_{OW} - 2.05$$

Assuming again that a power of 1.0 on K_{OW} applies results in the modified correlation for SXCF as

$$SXCF = 0.82 + 0.0065K_{OW}$$

If RCF and SXCF are assumed to be the equilibrium ratios of chemical concentrations in roots and stems to water and xylem sap, respectively, then

$$K_{RW} = Z_R\rho_W/(Z_W\rho_R)$$

and

$$K_{SXY} = Z_S \cdot \rho_{XY}/(Z_{XY} \cdot \rho_S)$$

where K_{RW} and K_{SXY} are root/water and stem/water equilibrium partition coefficients, respectively, and the ρ terms are densities of water (W), xylem sap (XY), root (R), and stem (S). The density correction is necessary since the correlations of Briggs et al. are on a mass fraction (g/g) basis whereas Z value are based on a mass/volume ratio (e.g., g/m³ or mol/m³).

In the absence of information to the contrary it is assumed that the xylem and phloem contents have the density and Z value of water, although it is recognized that the presence of dissolved organic matter could distort this relationship; thus,

$$Z_{Ph} = Z_{XY} = Z_W = 1/H$$

where H is the Henry's law constant.

The fugacity capacities of roots and stem become

$$Z_R = K_{RW} \cdot Z_w \cdot \rho_R/\rho_W = (0.82Z_W + 0.014Z_o) \, \rho_R/\rho_W$$

and

$$Z_S = K_{SXY} \cdot Z_X \cdot \rho_S/\rho_X = (0.82Z_W + 0.0065Z_o) \, \rho_S/\rho_W$$

The leaf is considered, for illustrative purposes, to consist of the equivalent of 18% air, 80% water, and 2% nonpolar organic matter and therefore has a fugacity capacity Z_L defined by

$$Z_L = 0.18Z_A + 0.8Z_W + 0.02Z_O$$

where Z_O is the Z value for octanol or $K_{OW}Z_W$. The minor amounts of other material, including mineral salts, are ignored in this calculation.

Lipid and water contents probably vary widely between species. Riederer (see Chapter 6 in this book), using a similar approach, suggests volume fractions of air, water, and lipids of 0.3, 0.648, and 0.005, respectively, for the European beech.

These tentative expressions for the fugacity capacities for the various plant parts, as summarized in Table 1, enable the deduction of Z values for a variety of chemicals from the physicochemical parameters of molecular weight, solubility, vapor pressure, and octanol/water partition coefficients as well as assumed water, air, and organic contents of the plant.

The bulk soil Z value, Z_B, is deduced, as discussed by Mackay and Paterson,[27] as the weighted sum of the Z values of the air (A), water (W), and organic carbon (OC) phases, namely

$$Z_B = 0.2Z_A + 0.3Z_W + 0.5\phi_{OC}Z_{OC}$$

where 0.2, 0.3, and 0.5 are the volume fractions of air, water, and solids, respectively, ϕ_{OC} is the organic carbon content of the solids, and Z_{OC} is the Z value for organic carbon, i.e., $K_{OC}Z_W$. A Z value for aerosol may be included as described by Mackay[24] using the aerosol-air partition coefficient K_{QA} given by

$$K_{QA} = 6 \times 10^6/P_L^S$$

hence

$$Z_Q = Z_A K_{QA}$$

where P_L^S is the subcooled liquid vapor pressure of the chemical.

C. TRANSPORT PROCESSES

Chemicals are transported through the plant by the xylem and the phloem. The xylem conducts the transpiration stream or sap from the roots to the stem and foliage. The phloem provides transport for organic matter from sites of photosynthesis to growing plant tissue. Chemicals may partition into, or react in or with, various plant tissues along the way or may be released to the atmosphere. To calculate D values within the plant, uniform flow rates, G_i ($m^3 \cdot h^{-1}$), are defined for flow in the xylem and phloem.

A transpiration flow rate of 5 $\mu g \cdot s^{-1}$ per square centimeter of foliage area (or 2×10^{-4} $m^3 \cdot h^{-1}$ per square centimeter of foliage) is adopted following Crank et al.[28] Since the total foliage area is 0.05 m^2, the resulting xylem flow rate G_{XY} is 1×10^{-5} $m^3 \cdot h^{-1}$, or 10 $cm^3 \cdot h^{-1}$, and the velocity in the xylem is 5 $m \cdot h^{-1}$. Flow rates in the phloem are less well quantified but are known to be slow compared to that of the xylem. The phloem flow rate (G_{Ph}) is assumed to be 5% of the xylem flow, or

$$G_{Ph} = G_{XY}/20 \ m^3 \cdot h^{-1}$$

Bulk flow in the xylem is described by the D parameter

$$D_{XY} = G_{XY} \cdot Z_W \ mol \cdot Pa^{-1} \cdot h^{-1}$$

and that in the phloem is defined by the equation

$$D_{Ph} = G_{Ph} \cdot Z_W \ mol \cdot Pa^{-1} \cdot h^{-1}$$

Organic chemicals in solution in soil water are assumed to enter the root by bulk flow and diffusion, with the former process being assumed dominant. The reverse process, from root to soil, is considered to be by diffusion only.

If uptake is simply through passive transfer of water and its associated dissolved chemical, the D value will be $G_{XY}Z_W$ as shown above. Processes such as (1) partitioning to root, stem, and foliage; (2) back diffusion; and (3) reflection at the endodermis may occur. The contribution of each of these processes to the reduction of chemical concentration in the transpiration stream is not fully understood. The model incorporates the first two processes. However,

if the chemical is retarded or reflected by a membrane barrier, the D value will
be reduced by some factor ϕ less than 1.0 to include this effect. Mc Farlane et
al.[29] have suggested that ϕ is 0.82 for uptake of nitrobenzene by the soybean.
This value will also include the effects of partitioning to plant tissue and
possible back diffusion. Active transport may give a ϕ exceeding 1.0. At this
stage it is assumed that the uptake D value is given by $\phi G_{XY} Z_W$ or ϕD_{XY} where
ϕ is 1.0. This is clearly an area in which more research is required to determine
the diffusive properties of the root.

The magnitude of the diffusive exchange D value from root to soil is
assumed to be equal to the flow in the phloem, i.e., 5% of the xylem flow. If
no term is included to permit chemical transfer from root to soil, absorption
becomes irreversible.

The D parameters for exchange between soil (subscript E) and root (subscript R) are thus

$$D_{ER} = \phi D_{XY} + D_{XY}/20 \text{ — soil to root (i.e., xylem flow and diffusion)}$$

$$D_{RE} = D_{XY}/20 \qquad \text{— root to soil (i.e., diffusion only)}$$

Transfer from root to stem is treated as an advective flow of sap in the
xylem, resulting in the parameter

$$D_{RS} = D_{XY} \text{ — root to stem}$$

The reverse process, from stem to root, occurs in the phloem and is described
by the parameter

$$D_{SR} = D_{Ph} \text{ — stem to root}$$

Exchange of chemical between stem and foliage is similarly assumed to occur
in the xylem and phloem. The resulting parameters for stem-foliage and
foliage-stem transfer of D_{SL} and D_{LS}, respectively, are

$$D_{SL} = D_{XY} \text{ — stem to foliage}$$

$$D_{LS} = D_{Ph} \text{ — foliage to stem}$$

Exchange of chemical between air and leaf occurs by diffusion through the
air boundary layer and then through the stomata into the interior of the leaf. It
also may occur by adsorption onto the cuticle with subsequent slower diffusion
through the cuticle. In chapter 6, Riederer gives a more detailed account of
these processes and suggests equations for quantifying these transport routes
among four compartments within the leaf. This approach is feasible and is
certainly preferred for a specific species, such as European beech, for which
extensive experimental data has been collected. At this stage the authors prefer

to treat the leaf as a single compartment and suggest an expression for a single leaf-air D value.

The numerical values for D_{LA} and D_{AL} are estimated from recent studies of leaf/air exchange kinetics of organic chemicals[14,21] in which the simple first-order equation

$$dC_L/dt = k_1C_A - k_2C_L$$

was applied, where C_L and C_A are the concentrations (mol $\cdot$ m^{-3}) in the leaf and air, respectively, and k_1 and k_2 are the uptake and clearance rate constants, respectively, with units of reciprocal hours. Chemical concentration in the leaf during uptake can be calculated as

$$C_L = C_Ak_1/k_2[1 - \exp(-k_2t)] = C_A \cdot BCF_V[1 - \exp(-k_2t)]$$

At infinite time, C_L/C_A becomes BCF_V (or Z_L/Z_A), the equilibrium ratio or bioconcentration factor of the volumetric chemical concentrations in leaf and air. BCF_V also can be expressed as k_1/k_2. During clearance

$$C_L = C_{LO} \exp(-k_2t)$$

where C_{LO} is the initial concentration in the leaf and C_A is zero. If the process is purely diffusive, the transfer parameters D_{LA} and D_{AL} are equal and the uptake-clearance equation becomes, for a leaf of volume V_L,

$$V_LZ_Ldf_L/dt = D_{LA}(f_A - f_L)$$

For uptake from an initial zero concentration, this equation can be integrated to give

$$f_L = f_A[(1 - \exp(-D_{LA} \cdot t/V_LZ_L)]$$

Clearly, k_2 is equivalent to D_{LA}/V_LZ_L or the reciprocal of the response time for leaf clearance.[30] The uptake rate constant k_1 is k_2BCF_V or k_2Z_L/Z_A or D_{LA}/V_LZ_A. The term D_{LA} can be viewed as the product of volume of air contacted per hour, G_{LA}, and Z_A; thus, k_1 is G_{LA}/V_L or the number of leaf volumes of air contacted per hour. The leaf uptake data suggest a maximum value of k_1 of about 20,000 h^{-1}; thus, apparently each leaf contacts 20,000 times its volume of air per hour.

Now if it is assumed that D_{LA}, the overall conductivity, can be calculated from the overall resistance $1/D_{LA}$, which comprises series resistances in air ($1/D_A$) and in cuticle ($1/D_C$), then

$$1/D_{LA} = 1/D_A + 1/D_C$$

and

$$\frac{1}{k_2} = \frac{V_L Z_L}{D_{LA}} = \frac{V_L Z_L}{D_C} + \frac{V_L Z_L}{D_A}$$

Using experimental values for the clearance rate constant from the work of Bacci et al.,[14,21] Paterson et al.[30] correlated k_2 with physicochemical properties for 14 organic compounds with the similar relationship

$$1/k_2 = \tau_0 + \tau_A K_{OA}$$

where K_{OA} is the octanol/air partition coefficient; τ_0 and τ_A represent plant-specific transfer times or resistances in the organic (or leaf cuticle) and air phases and have estimated values of 126 and 5×10^{-6} hours, respectively. Values of D_C, D_A, and D_{LA} (and D_{AL}) can be calculated for each chemical as follows:

$$D_C = V_L Z_L / \tau_O$$

$$D_A = V_L Z_L / (\tau_A K_{OA}) = V_L Z_L \cdot Z_A / (\tau_A Z_O) \approx 0.05 V_L Z_A / \tau_A$$

since $Z_L \approx 0.05 Z_O$

$$D_{LA} = D_{AL} = 1 / (1/D_C + 1/D_A)$$

The soil is assumed to be 5 cm deep and of horizontal area A m², containing 20% (by volume) air and 30% water. The balance is mineral matter containing 0.15% organic carbon. Diffusion from soils can take place in the air or water phase or in parallel in both. D values for this process have been developed previously[24,27] following the approach used by Jury et al.[31]

The D values for parallel diffusive flows in air and water in the soil are expressed as

$$D_{AD} = k_A A Z_A$$

$$D_{WD} = k_W A Z_W$$

where k_A and k_W can be considered as mass transfer coefficients or transport velocities in the soil, air and water, respectively. Respective values of 0.015 and 6×10^{-6} m · h^{-1} are calculated as the quotient of a typical effective diffusivity in the appropriate phase (air and water) and at the soil diffusion depth of 2.5 cm, or half the total depth.

The air-side boundary layer diffusion parameter consists of a mass transfer coefficient or transport velocity, k_B (1 m · h^{-1}), and is

$$D_{SA} = k_B A Z_A$$

The total D value for soil-air diffusion becomes

$$D_{EA} = 1/[1/D_{SA} + 1/(D_{AD} + D_{WD})]$$

It is possible to include nondiffusive processes of air-to-soil deposition by wet and dry deposition of aerosol-associated chemical. For chemicals of low vapor pressure which partition appreciably to aerosols these processes can be more important than diffusive adsorption. Methods of estimating these D values are described by Mackay (1991). The total value for air to soil transfer is then

$$D_{AE} = D_{EA} + D_{QR} + D_{DP}$$

where D_{QR}, the parameter for chemical transfer by rain, is the product of the effective rain velocity (typically 1×10^{-4} m^3 m^{-2} h^{-1} or $m \cdot h^{-1}$), the soil area A m^2, and Z_W; D_{DP}, the parameter for aerosol-associated chemical transfer, is the product of the "effective deposition velocity" (typically 6×10^{-10} $m \cdot h^{-1}$, the soil area, and Z_Q, the aerosol Z value. This effective deposition velocity is defined as the product of the actual deposition velocity and the volume fraction of aerosols in the air.

Inflow of air to the system is assumed to occur at a flow rate G_{BA} $m^3 \cdot h^{-1}$ which can be estimated from an actual flow rate in controlled laboratory conditions or from wind speed under environmental conditions. Influx of chemical of concentration C_{BA} $mol \cdot m^{-3}$ thus occurs at a rate

$$N_{BA} = G_{BA} C_{BA} = G_{BA} f_{BA} Z_A = D_{BA} f_{BA} \quad mol \cdot h^{-1}$$

and D_{BA} is $G_{BA} Z_A$.

D. TRANSFORMATION PROCESSES

For each compartment i, a metabolism rate D_{iM} can be included as $V_i Z_i k_{Ri}$, where k_{Ri} is the first-order metabolic rate constant (h^{-1}). Other degrading reactions such as hydrolysis or photolysis may be included in an overall k_{Ri}. It is often convenient to express k_{Ri} as a half-life.

These D values for transport and transformation are illustrated in Figure 1, and the D value expressions are summarized in Table 3.

E. SUMMARY

At this stage tentative expressions have been suggested for appropriate partitioning, transport, and transformation expressions which are applicable to any chemical in this illustrative system. The primary aim has been to establish

Table 3.　Summary of expressions for D values

Transfer between	D values	
Earth-root	$D_{ER} = \phi D_{XY} + D_{XY}/20$	Xylem flow and diffusion
Root-earth	$D_{RE} = D_{XY}/20$	Diffusion only
Root-stem	$D_{RS} = D_{XY}$	Xylem flow
Stem-root	$D_{SR} = D_{Ph}$	Phloem flow
Stem-foliage	$D_{SL} = D_{XY}$	Xylem flow
Foliage-stem	$D_{LS} = D_{Ph}$	Phloem flow
Foliage-air	$D_{LA} = 1/(1/D_C + 1/D_A)$	
	$D_C = V_L Z_L/\tau_O$	Diffusion through cuticle
	$D_A = V_L Z_L/(\tau_A K_{OA})$	Diffusion through air boundary layer
Air-foliage	$D_{AL} = D_{LA}$	
Air-soil	$D_{AE} = 1/(1/D_{SA} + 1/(D_{AD} + D_{WD})) + D_{QR} + D_{DP}$	
	$D_{SA} = K_B A Z_A$	Diffusion through air boundary layer
	$D_{AD} = K_A A Z_A$	Diffusion in soil air
	$D_{WD} = K_W A Z_W$	Diffusion in soil water
	$D_{QR} = 1.14 \times 10^{-4} A Z_W$	Rain rate
	$D_{DP} = 6.0 \times 10^{-10} A Z_Q$	Particle deposition
Soil-air	$D_{EA} = 1/(1/D_{SA} + 1/(D_{AD} + 1/D_{WD}))$	

reasonable expressions and suggest approximate values for the parameters. Undoubtedly as more information emerges on chemical fate in such systems, the expressions and parameters will be improved, and this information can be incorporated into the model. Clearly the expressions for root-soil exchange and leaf-air exchange require refining. In no sense are the expressions suggested here regarded as final.

IV. MODEL INTERPRETATION

Having established estimates of volumes (V), Z values, and D values, it is useful to gather them together in a process diagram, as illustrated in Figures 2 and 3, which are for hexachlorobenzene (HCB) and dichlorobenzonitrile (DCBN) in a 1 m³ growth chamber similar to that developed by Bacci et al.[21] and containing a plant similar to soybean. A value of approximately 100 h was selected for τ_o as being representative of soybean. The properties of these chemicals are given in Table 4. No reactions or metabolism are included at this stage. In the interests of simplicity, aerosols and aerosol transport are also not included.

There are four quantities which are of interest. First, the values of Z show where concentrations are likely to be highest because at a constant prevailing fugacity f, C equals Zf. In this case, the Z values (mol · m^{-3} · Pa^{-1}) for HCB are leaves, 48.3; roots, 28.2; stem, 13.0; and bulk soil, 1.86. Air is 4×10^{-4}. It is thus likely that HCB will be most readily detected in leaf tissue. For DCBN leaf also is the highest (28), followed by roots (16), stem (8.2), soil (1.52), and air (4×10^{-4}).

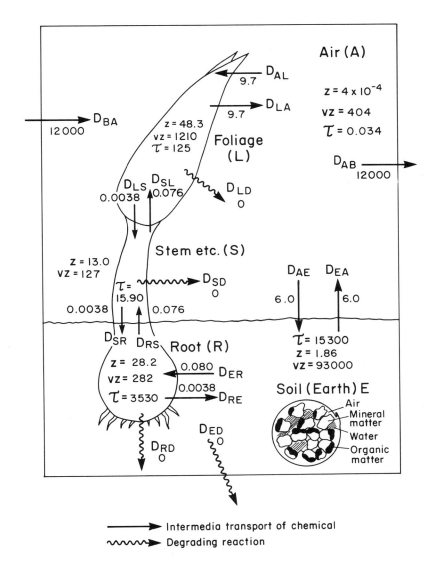

FIGURE 2. Diagram of soil-plant-air system showing Z values, VZ values ($\times 10^6$), D values ($\times 10^6$), and response times (τ, h) for hexachlorobenzene.

Second, values of the VZ product for each compartment show where the capacity for chemical is largest, i.e., where most of the chemical is likely to be located. In the case of HCB, the soil VZ (0.093) greatly exceeds that of the root (0.00028), the leaves (0.0012), and the stem (0.00013). The air VZ is negligible (0.0004). In such a system over 99% of the chemical is likely to remain in the soil, with perhaps 1.3% in foliage and 0.3% in the root. Clearly the capacity of

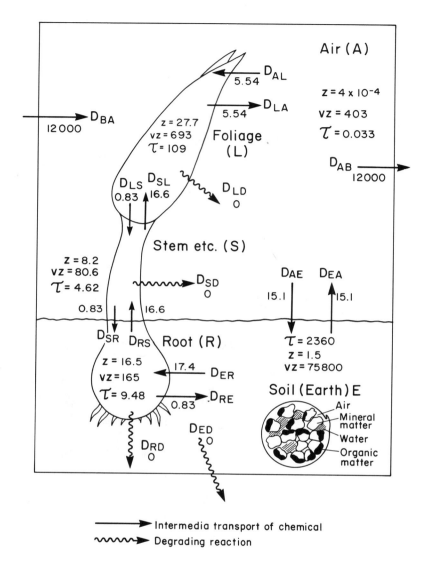

FIGURE 3. Diagram of soil-plant-air system showing Z values, VZ values ($\times 10^6$), D values ($\times 10^6$), and response times (τ, h) for dichlorobenzonitrile.

the plant is negligible compared to that of the soil; thus, it cannot appreciably deplete chemical from the soil. For DCBN the soil VZ is 0.076, which also greatly exceeds that of the root (1.65×10^{-4}), the leaves (6.9×10^{-4}), and the stem (8×10^{-5}). Again only about 1% will be found in the plant tissues.

Table 4. Physicochemical properties of hexachlorobenzene (HCB)[32] and dichlorobenzonitrile (DCBN)[33]

Property (units)	HCB	DCBN
Molecular weight (g/mol)	284.79	172.02
Aqueous solubility (g/m³)	0.005	20
Vapor pressure (Pa)	0.0023	0.07
Log octanol/water partition coefficient	5.5	2.9

The D values for transport and transformation are the third point of interest. A large D value implies a high conductivity and, thus, potentially rapid transport or transformation. Inspection of the magnitudes of the D values indicates which processes are potentially most important, especially when there are parallel routes of transport.

The most striking feature of the D values in both Figures 2 and 3 is that the advective D values are over a factor of 1000 greater than the other process D values. The implication is that the concentration in the air will be dominated by advective flow, i.e., dilution, and only a small fraction of the chemical in the air will be removed by the foliage.

In this case the D values for soil-to-leaf transport of HCB are: evaporation (6.0×10^{-6}) followed by absorption to foliage (9.7×10^{-6}) or, alternatively, transfer from soil to root (0.080×10^{-6}), root to stem (0.076×10^{-6}), and stem to leaf (0.076×10^{-6}) suggesting that the evaporation route will be easier by about a factor of 80. It is likely that the HCB concentration in the foliage will be controlled primarily by how fast the chemical can evaporate into the air space. That there is quite rapid and intimate contact between foliage and the air is evident from the high exchange D values. A more hydrophobic chemical such as a PCB which has a lower water solubility will experience even more difficult transport through the plant; thus, leaves will be contaminated almost entirely by evaporation from soil.

For DCBN, transfer through the plant is easier, with a series of three D values of about 17×10^{-6}, whereas evaporation and foliar absorption D values are 15.1×10^{-6} and 5.5×10^{-6}, respectively. A more water soluble and less volatile substance such as nitrobenzene or 2,4-dichlorophenoxyacetic acid (2,4-D) will reach leaves almost entirely by the root-stem-foliage route.

If metabolic conversion D values had been included, the relative rates of transport and transformation would become obvious immediately.

The fourth point of interest is the time constant τ hours for each phase as influenced by each process. This can be deduced, as is discussed later, as VZ/D, where D is the process D value and VZ is for the compartment in question. It is approximately the time required for the phase to respond to a change in concentration as a result of transport from adjacent phases or transformation in the phase. Selected times are also shown in Figures 2 and 3. Inspection of the differential equations shows that for a compartment, such as the stem, the time of response is VZ divided by the sum of the D values for <u>loss</u> from the stem.

These response times also can be estimated from the quantity of chemical in the compartment at steady-state divided by the rate of chemical input or output.

The τ value for the air phase for both chemicals is very short, being 0.034 h, or 2 min, for both chemicals. This is because the air space is flushed rapidly. The air will thus come to steady-state within a few minutes and will maintain a constant, low concentration controlled by the relative rates of evaporation and advective loss.

In the case of HCB the leaf τ for exchange with air is about 125 h; thus, the leaf will approach to within 37% or $\exp(-1)$ of its steady-state value with respect to the air in 5 days. For evaporation from soil τ is 15,500 h, or 1.8 years; thus, there will be negligible loss of HCB by evaporation or by root uptake in any reasonable time period. For uptake by roots from soil, τ is 3500 h, or 147 days; thus, in the lifetime of the plant it can barely approach equilibrium with the soil. This is because the very low Z value in the water phase constrains the root uptake D value to a very low value. It is possible that the outer surface of the root will be contaminated as a result of receiving chemical by diffusion through air pores and will rapidly achieve equilibrium with the soil. However, it seems unlikely that the bulk of the root, which is 80% water, will be affected. Transport from root to leaf is even less likely.

It thus appears that during the time course of an experiment, which may last 1000 h, the most readily detectable response will be uptake of HCB by leaves from air as a result of evaporation from soil. The soil concentration will remain fairly constant, and only the outer surface of the root will be contaminated. These deductions are invaluable because they guide the solution of the mass balance equations discussed in Section V.

In the case of DCBN, the τ for foliar exchange with air is 126 h, or 5.3 days. The τ for loss of chemical from soil is 5000 h, or 210 days, while the τ for air receiving chemical from the soil is 27 h. The τ for root uptake is a fairly short 9.5 h. It is thus likely that transport into the root will be completed fairly rapidly, within a few days, with transport to air and leaf being slower. The first chemical to reach the leaves will likely be by transpiration. It is notable that Mc Farlane et al.[29] showed that DCBN transport from hydroponic solution had a half-time of about 10 h, in qualitative agreement with these assertions.

V. MASS BALANCE EQUATIONS AND SOLUTIONS

Having assembled information concerning the Z, VZ, and D values, the next stage is to write the mass balance equations in differential form, then explore how they can be solved to yield solutions which can be reconciled with experimental data, and ultimately build up a predictive capability from successful modeling simulations.

For each compartment (air, soil, root, stem, and foliage) a differential equation of the form

$$Z_i d(V_i f_i)/dt = E_i + \Sigma D_{ji} f_j - f_i (\Sigma D_{ij} + D_{iM})$$

can be written. Thus, assuming the Z values to be constant with time,

$$Z_i V_i df_i/dt + Z_i f_i dV_i/dt = E_i + \Sigma D_{ji} f_j - f_i D_{iT}$$

The accumulation term on the left contains the phase volume V, bulk Z value, fugacity f, and the derivatives of the volume and fugacity with respect to time. The term dV_i/dt is the phase growth rate and is applicable only to plant compartments. Often it may be ignored. The term $Z_i dV_i/dt$ can be regarded as a "growth dilution" D value, D_{iG}. If it is fairly constant, it can be taken to the right side of the equation and incorporated with the other removal D values. E_i is the emission rate to the phase (mol · h^{-1}). The group $\Sigma D_{ji} f_j$ is the sum of the transport rates from all other media to the phase, while the group $\Sigma D_{ij} f_i$ is the sum of the transport rates from the phase. The group $f_i D_{iM}$ is the rate of loss of chemical by reaction or metabolism. It is convenient to group all the loss terms in one group, $f_i D_{iT}$, where $D_{iT} = \Sigma D_{ij} + D_{iM}$, with D_{iG} possibly being included.

The most direct and general solution strategy is to establish initial conditions in all five phases, then calculate, by numerical integration, the changing fugacities as they respond to transport, transformation, and variations in emission rate. Plant growth can be included if desired. The problem with this approach is that it lacks generality; i.e., each solution is specific to the initial conditions and assumed emissions. Solution also may be quite slow because some fast processes will require small integration time steps.

The authors of this chapter suggest that it is advantageous to simplify the equations and reduce them where possible to algebraic form by making one or more "steady-state" assumptions. As discussed earlier, a time constant τ can be estimated for each phase as approximately the time required to approach to (1 − e^{-1}) or 63% of the steady-state value with respect to adjacent phases. If the time frame of interest in the overall calculation is a large multiple of τ, i.e., 10 or more, then it is a fair assumption that steady state applies, the differential term on the left can be set to zero, and the equation can be reduced to algebraic form. This simplifies and shortens the calculation considerably. This can be done for one or more phases depending on the objective of the calculation and the properties of the chemical and the system.

It is also interesting to reduce all the equations to algebraic form and examine the full steady-state solution. This solution may or may not have physical significance because the time required to reach steady state may exceed the life-span of the plant or the period of experimental interest, but it does provide an indication of the ultimate partitioning characteristics of the system. It is especially valuable when this solution is examined in the light of the time constant terms τ_i for each medium. If τ_i is of magnitude similar to the period of interest, or is longer, this indicates that the system is under "kinetic control", and equilibrium or steady-state calculations are not meaningful.

Figure 4 depicts a steady-state mass balance for DCBN which could be reached after approximately 7 months. This period could represent one growing season in some climates. Mc Farlane et al.[29] noted that DCBN was recovered entirely as parent compound in the roots of a soybean plant, but only about 50% was recovered as parent compound in the leaves. Assuming a metabolic half-life of 100 h in the foliage, which is similar in magnitude to the time constant for that compartment, results in a metabolic rate constant and D value of approximately 0.007 h^{-1} and 4.85×10^{-6} mol $\cdot$ Pa^{-1} $\cdot$ h^{-1}, respectively. By assuming a soil concentration of 1 g $\cdot$ m^{-3} and a foliar metabolic rate constant of 0.007 h^{-1} and applying an algebraic solution, the steady-state concentrations in the various compartments can be calculated. The foliar time constant is reduced to approximately 60 h.

As Figure 4 shows, the primary transport process is uptake of chemical by the root from the soil with subsequent translocation. Although a comparable amount evaporates from the soil, most of this is destined to be advected from the system; thus, there is little absorption by foliage from the air. About half the DCBN entering the leaves evaporates and half is metabolized.

Finally, Figure 5 shows the calculated unsteady-state response of the plant tissues to constant soil and air concentrations of DCBN, identical to those shown in Figure 4. The root concentration rises from zero to 14.3 μg $\cdot$ g^{-1}, reaching 7 μg $\cdot$ g^{-1} (or half the final value) in about 6 h. The stem concentration rises from zero to 7.4 μg $\cdot$ g^{-1}, reaching 3.7 μg $\cdot$ g^{-1} in about 12 h. The foliage is slower, taking 61 h to reach 18.5 μg $\cdot$ g^{-1}, or half its final value of 37 μg $\cdot$ g^{-1}.

Results such as these can be compared directly with experimental data obtained in experiments such as those described by Mc Farlane et al.[29] This provides a rigorous test of the model assumptions and parameters. A detailed comparison of experimental and model results is beyond the scope of this chapter, but the authors believe that a model such as this should be capable of being usefully applied to such data.

VI. CONCLUSIONS

A multicompartment approach for a plant-air-soil system has been developed and a compartment segmentation scheme suggested. However, it is not viewed as the only, or even the best, segmentation. Models such as this require information on equilibrium partitioning and kinetics. The use of the fugacity/ Z value approach for expressing equilibrium in terms of actual or pseudo contents of air, water, and octanol for various plant tissues appears to be adequate. More data are needed for a variety of chemicals and tissues to strengthen the correlations; thus, the equations suggested here should be viewed as tentative in nature and subject to modification. Expressing the kinetic transport and transformation processes in terms of D values is also regarded as

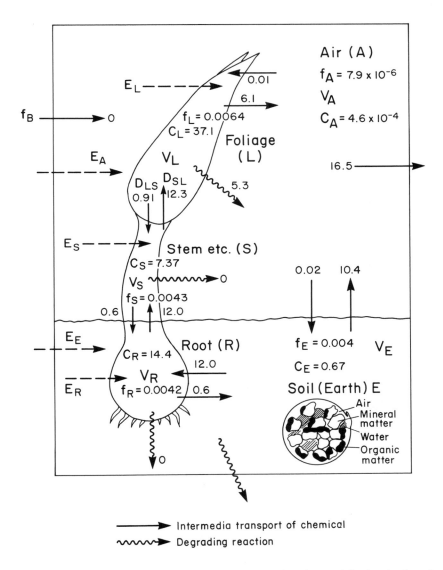

FIGURE 4. Illustrative steady-state model output for dichlorobenzonitrile showing fugacities (fPa), concentrations (C, μg/g), and transport and metabolism rates (μg/h).

feasible and adequate, but again more data are needed to express accurately rates of transport to the plant and within the plant. Soil-to-root transport and phloem transport are particularly important and poorly understood.

It is suggested that inspection of the Z values, the VZ capacity groups, the D values, and the VZ/D response times provides valuable insights into

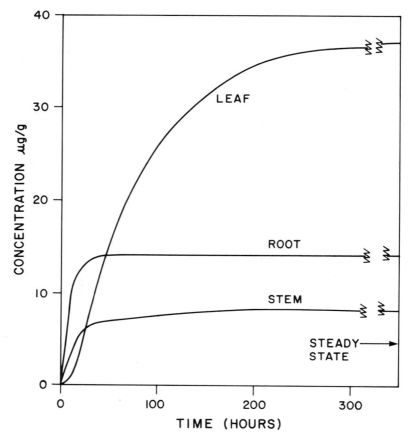

FIGURE 5. Plot of the time course of concentration increases of dichlorobenzonitrile in root, stem, and foliage as they approach the steady-state values in Figure 4.

contaminant behavior in the soil-plant-air system. The readily obtained algebraic steady-state solution provides a valuable indication of chemical behavior. Assembling the multicompartment differential equations is relatively straightforward, and solution by numerical integration is fairly routine. These solutions, however, lack generality and do not always lead to an intuitive appreciation or understanding of the key processes.

What are clearly needed are experimental data for a number of chemicals of known properties, gathered with a view to fitting the results to a model such as this. The resulting experiences will be valuable for exposing weaknesses in parameter values and estimation methods and for suggesting structural improvements for the model. Ultimately, it is likely that a number of validated plant models will emerge which can reliably describe the partitioning, transport, and transformation of chemicals under a variety of exposure conditions.

Such a capability will be invaluable for solving the broader problem of quantifying chemical migration in the biosphere of which we are a part.

REFERENCES

1. Calamari, D., Bacci, E., Focardi, S., Gaggi, C., Morosini, M., and Vighi, M. (1991), Role of plant biomass in the global environmental partitioning of chlorinated hydrocarbon, *Environ. Sci. Technol.*, 25: 1489.
2. Smillie, R.H. and Waid, J.S. (1986), Australia wide monitoring of airborne polychlorinated biphenyls (PCBs) using Eucalyptus foliage, *Search,* 17: 216.
3. Gaggi, C., Bacci, E., Calamari, D., and Fanelli, R. (1985), Chlorinated hydrocarbons in plant foliage: An indication of tropospheric contamination level, *Chemosphere,* 14: 1673.
4. Bacci, E., Calamari, D., Gaggi, C., Biney, C., Focardi, S., and Morosini, M. (1988), Organochlorine pesticide and PCB residues in plant foliage (*Mangifera indica*) from West Africa, *Chemosphere,* 17: 693–702.
5. Hinkel, M., Reischl, A., Schramm, K.-W., Trautner, F., Reissinger, M., and Hutzinger, O. (1989), Concentration levels of nitrated phenols in conifer needles, *Chemosphere,* 18: 2433.
6. Reischl, A., Reissinger, M., and Hutzinger, O. (1987), Occurence and distribution of atmospheric organic micropollutants in pine needles, *Chemosphere,* 16: 2647.
7. Jensen, S., Eriksson, G., Kylin, H., and Strachan, W.M.J. (1992), Atmospheric pollution by persistent organic compounds monitoring with pine needles, *Chemosphere,* 24: 229.
8. Hermanson, M.H. and Hites, R.A. (1990), Polychlorinated biphenyls in tree bark, *Environ. Sci. Technol.*, 24: 666.
9. Scheunert, I. (1985), Prediction of chemical's behavior in soil-plant systems from physicochemical properties, in "Environmental Modeling for Priority Setting among Existing Chemicals," Proceedings of GSF Workshop, Munich, Nov. 11–13, 1985, pp. 279–298.
10. Topp, E., Scheunert, I., Attar, A., and Korte, F. (1986), Factors affecting the uptake of ^{14}C labeled organic chemicals by plants from soil, *Ecotoxicol. Environ. Saf.,* 11: 219.
11. Briggs, G.G., Bromilow, R.H., and Evans, A.A. (1982), Relationships between lipophilicity and root uptake and translocation of non-ionized chemicals in barley, *Pestic. Sci.,* 13: 495.
12. Villeneuve, J.P. and Holm, E. (1984), Atmospheric background of chlorinated hydrocarbons studied in Swedish lichens, *Chemosphere,* 13: 1133.
13. Travis, C.C. and Hattemer-Frey, H.A. (1988), Uptake of organics by aerial plant parts: A call for research, *Chemosphere,* 17: 277.
14. Bacci, E., Calamari, D., Gaggi, C., and Vighi, M. (1990), Bioconcentration of organic chemical vapors in plant leaves: Experimental measurements and correlation, *Environ. Sci. Technol.*, 24: 885–889.
15. Trapp, S., Matthies, M., Scheunert, I., and Topp, E. (1990), Modeling the bioconcentration of organic chemicals in plants, *Environ. Sci. Technol.*, 24: 1246.

16. Ryan, J.A., Bell, R.M., Davidson, J.M., and O'Connor, G.A. (1989), Uptake of nonionic organic chemicals from soils, *Chemosphere,* 17: 2299.

17. Hawxby, K., Basler, E., and Santelmann, P.W. (1972), Temperature effects on absorption and translocation of Trifluralin and Methazole in peanuts, *Weed Sci.,* 20: 285.

18. Paterson, S., Mackay, D., and Gladman, A. (1991), A fugacity model of chemical uptake by plants from soil and air, *Chemosphere,* 23: 539.

19. Calamari, D., Vighi, M., and Bacci, E. (1987), The use of terrestrial plant biomass as a parameter in the fugacity model, *Chemosphere,* 16: 2359.

20. Schramm, K.W., Reischl, A., and Hutzinger, O. (1987), A multimedia compartment_model to estimate the fate of lipophilic compounds in plants, *Chemosphere,* 16: 2653.

21. Bacci, E., Cerejeira, M.J., Gaggi, C., Chemello, G., Calamari, D., and Vighi, M. (1990), Bioconcentration of organic chemical vapors in plant leaves: The azalea model, *Chemosphere,* 21: 525–535.

22. Riederer, M. (1990), Estimating partitioning and transport of organic chemicals in the foliage/atmosphere system: Discussion of a fugacity based model, *Environ. Sci. Technol.,* 24: 829.

23. Mackay, D. and Paterson, S. (1982), Finding fugacity feasible, *Environ. Sci. Technol.,* 16: 654A.

24. Mackay, D. (1991), *Multimedia Environmental Modeling: The Fugacity Approach,* Lewis Publishers, Boca Raton, FL.

25. Briggs, G.G., Bromilow, R.H., Evans, A.A., and Williams, M. (1983), Relationships between lipophilicity and the distribution of non-ionised chemicals in barley shoots following uptake by roots, *Pestic. Sci.,* 14: 492.

26. Paterson, S. and Mackay, D. (1989), Modeling the uptake and distribution of organic chemicals in plants in *Intermedia Pollutant Transport: Modeling and Field Measurements,* D.T. Allen, Y. Cohen, and I.R. Kaplan, Eds., Plenum Press, New York, pp 269–281.

27. Mackay, D. and Paterson, S. (1991), Evaluating the multimedia fate of organic chemicals: A level III fugacity model, *Environ. Sci. Technol.,* 25: 427.

28. Crank, J., Mc Farlane, N.R., Newby, J.C., Paterson, G.D., and Pedley, J.B. (1981), *Diffusion Processes in Environmental Systems,* Macmillan, London.

29. Mc Farlane, C., Nolt, C., Wickliff, C., Pfleeger, T., Shimabuku, R., and McDowell, M. (1987), *Environ. Toxicol. Chem.,* 6: 847.

30. Paterson, S., Mackay, D., Bacci, E., and Calamari, D. (1991), Correlation of the equilibrium and kinetics of leaf-air exchange of hydrophobic organic chemicals, *Environ. Sci. Technol.,* 25: 866.

31. Jury, W.A., Spencer, W.F., and Farmer, W.J. (1983), Behavior assessment model for trace organics in soil. I. Model description, *J. Environ. Qual.,* 12: 558.

32. Mackay, D., Shiu, W.Y., and Ma, K.C. (1992), *Illustrated Handbook of Physical-Chemical Properties and Environmental Fate for Organic Chemicals,* Vol. 1, Lewis Publishers, Boca Raton, FL.

33. Suntio, L.R., Shiu, W.Y., Mackay, D., Seiber, J.N., and Glotfelty, D. (1988), Critical review of Henry's law constants for pesticides, *Rev. Environ. Contam.,* 103: 1.

CHAPTER **8**

Dynamics of Leaching, Uptake, and Translocation: The Simulation Model Network Atmosphere-Plant-Soil (SNAPS)

Michael Matthies and Herwart Behrendt

TABLE OF CONTENTS

I. INTRODUCTION

The behavior of chemicals in soils, their uptake into plants, and their exchange with the atmosphere and groundwater are governed by the water dynamics in soil and plant during the vegetation period. Soil water flux, adsorption/desorption to soil particles, and degradation determine chemical transport and fate during the soil passage. Substances with intermediate water solubility and moderate lipophilicity can be systemically taken up by roots and translocated into stem, leaves, and fruits. The dynamics and quantities of uptake depend on the dissolved concentrations at the soil depth where the root uptake is most effective. In this chapter a system of simulation models called SNAPS (Simulation Model Network Atmosphere-Plant-Soil) is described which combines fate models for chemicals in soil and plant and their exchange with the atmosphere. Transport, uptake, and translocation of three pesticides are simulated in two different scenarios.

II. MODEL DESCRIPTION

A. SOIL WATER MODEL

The soil water model is based on the model SWACRO.[1] The model solves the one-dimensional soil water transport equation (Richard's equation) to calculate the updated soil water content. The equation is

$$\frac{\partial \Psi(z,t)}{\partial t} = \frac{1}{C_{apw}(\Psi)} \frac{\partial\left[K(\Psi)\left(\frac{\partial\Psi}{\partial z}+1\right)\right]}{\partial z} - \frac{S(z,t)}{C_{apw}(\Psi)} \tag{1}$$

where t   = time (days),
       z   = depth in soil (cm),
       $\Psi(z,t)$ = soil matrix potential (cm H_2O),
       C_{apw} = differential water capacity

$$= \frac{\partial\theta}{\partial\Psi}(1/cm\ H_2O),$$

       θ = vol. water content (–),
       K   = soil hydraulic conductivity (cm/day),
       $S(z,t)$ = plant water uptake rate (cm H_2O/cm/day).

As the boundary condition at the soil surface the evapotranspiration is calculated after Monteith and Rijtema.[1] Free drainage is assumed at the lower boundary at 2m depth. The root water uptake has been modified so that the potential uptake is determined by the root length distribution in soil. In the case of water stress the potential uptake is reduced by an empirical function.[1] The root length distributions of cereals are determined by a regression equation to soil texture and phenomenological stage (shooting, booting, flowering, and ripening).[2] The sink term for root water uptake is then

$$S(z,t) = \alpha[\Psi(z,t)]T_p(t)\frac{w(z,t)}{\displaystyle\int_0^{z_{root}} w(z,t)dz}$$

(2)

where $T_p(t)$ = potential transpiration (cm H_2O /day),
  $w(z,t)$ = root length distribution (cm/cm^3),
  $\alpha(\Psi)$ = empirical reduction function in case of water stress (–),
  z_{root} = rooting depth (cm).

B. SOIL CHEMICAL TRANSPORT MODEL

The soil chemical transport model is based on the convection-dispersion equation. Details for the chemical transport may be found elsewhere.[3] The model assumes equilibrium distribution between dissolved and sorbed phases. The upper boundary condition is specified as a time-constant chemical input pulse. Because of the low vapor pressure of the chemicals used in the scenarios below, transport in the gaseous phase and, thus, volatilization from soil to atmosphere is neglected. The equation is

$$\frac{\partial(\theta C + \rho K_d C)}{\partial t} = \frac{\partial\left(D(\theta,q)\dfrac{\partial C}{\partial z}\right)}{\partial z} - \frac{\partial(qC)}{\partial z} - S_c$$

(3)

where C = dissolved concentration (μg/cm^3),
  D (ϕ,q) = effective diffusion-dispersion coefficient (cm^2/day),
  S_c = chemical sink term (degradation, root uptake; μg/cm^3/day),
  K_d = linear equilibrium adsorption coefficient (K_d = OC K_{oc}; cm^3/g),
  q = Darcy water velocity in soil (cm/day),
  ρ = soil bulk density (g/cm^3).

The sink term S_c in Equation 3 describes the biotic and abiotic degradation in soil and the root uptake with transpiration water:

$$S_c = S_{deg}(z,t) + S_{root}(z,t)$$

(4)

The degradation in soil is described by a bulk first-order rate equation:

$$S_{deg}(z,t) = g(T_s)\ f(z)\ \mu_0\ C_T(z,t)$$

(5)

with C_T = total chemical concentration in soil (μg/cm^3);
      μ_0 = bulk first-order degradation rate at 20°C in upper soil horizon
              (1/day);
      $f(z)$ = empirical depth-in-soil function; i.e., reduction of degradation
              rates in deeper soil layers (–);
      $g(T_s)$ = temperature correction by van't Hoff's rule (–);
      T_s = temperature in soil (°C).

Since soil degradation is significantly influenced by soil temperature, the model calculates the temperature by a numerical solution of a diffusive-type heat transport equation. Measured soil surface temperatures are used as boundary conditions. The calculations follow the model of Campbell.[4]

The chemical root uptake is calculated via a mass flux term and the transpiration stream concentration factor (TSCF).[5] The TSCF describes the distribution of the organic chemical between soil water and crop transpiration stream. The TSCF is closely correlated to the lipophilicity (n-octanol/water partition coefficient K_{ow}) of the chemical. Equation 7 is an empirical equation found for barley.[5] The chemical uptake has a pronounced optimum for log K_{ow} = 1.78.

$$S_{root}(z,t) = TSCF \ S(z,t) \ C(z,t) \tag{6}$$

$$TSCF = 0.784 \ exp[-(\log K_{ow} - 1,78)^2/2.44] \tag{7}$$

C. PLANT FATE MODEL

The chemical fate within the plant as translocation to stem, leaves, and fruits, metabolization, and exchange with the atmosphere is calculated with the plant model described in Reference 6.

III. DATA REQUIREMENT AND DATA SUPPORT FOR SCENARIO ANALYSIS

Two different scenarios were chosen to analyze the uptake behavior of three water-soluble pesticides. The data for the two scenarios were provided from a 3-year project on modeling the transport in the field under normal agricultural practices. The aim of the project was to evaluate the predictive power of soil leaching models if they were based solely on public survey data sets (soil information system, climate and crop practices, substance properties). Pesticide residues were not measured in the crop; thus, all simulated concentrations could not be compared to measured values. The results show the general principles of the simultaneous leaching, systemic uptake, and translocation of pesticides over a vegetation period of cereals.

A. SOIL PROPERTIES

The fields studied are located in eastern Bavaria. A sandy soil (dystric cambisol) under winter barley near Neumarkt/Oberpfalz and a loamy soil (luvisol) under winter wheat near Regensburg were selected. The distance between the two locations is about 100 km. They represent typical agricultural soils with different climate regimes. The physical data of the selected soils were measured in layers of 25 cm thickness and are given in Table 1. Three "horizons" (I, II, III) were defined which differ in their physical and chemical properties. The upper horizon (I) is the Ap horizon; the second horizon (II) is 50 cm deep for the sandy soil and 25 cm deep for the loamy soil. The third horizon (III) ranges to 200 cm with free drainage of water (no interaction with groundwater). Table 1 shows the soil properties that are important for the behavior of organic chemicals in soil. In particular, the organic carbon contents and the field capacities are quite different for the two soils.

B. AGRICULTURAL PRACTICE AND CLIMATE DATA

The main part of the vegetation period was simulated starting from the application of the pesticides on April 10, 1990 and ending with the harvest (Table 2). The application of the pesticides was assumed to be on the soil surface. Winter barley was harvested on July 11 and winter wheat on August 5. The simulation interval was therefore 92 days for winter barley and 117 days for winter wheat. The time series of precipitation and air temperature are given in Figure 1 for both scenarios. (Julian days are counted from January 1). The climate data were taken from the next agrometeorological station (about 20 km away). Temperature was similar at both locations. However, precipitation showed quite different time patterns. The highest intensities were observed in Neumarkt (sand) during the first 20 days of the simulation period and after about 2 months. Rainfall was highest in Regensburg (loam) during the third month of the simulation period, with a single peak just before harvest. Total precipitation was 83 mm for Neumarkt (sand) and 178 mm for Regensburg (loam). These different precipitation intensities and patterns and the different soil properties were responsible for the much lower yield of barley compared to wheat. Table 2 summarizes the important data for both scenarios.

C. CROP PROPERTIES

The modeling and simulation of the uptake of chemicals into plants require a set of data on crop properties (see Table 3). Both crops are cereals. Lipid and water contents in stem, leaves, and corn are quite similar for both crops. Their growth and development differ mainly because of the water and nutrient supply. At the time of the pesticide application (April 4, 1990), a standing fresh biomass (stem and leaves) of 768 kg/ha for winter barley and of 824 kg/ha for winter wheat was estimated.

Table 1. Physical properties of sand and loam soil

Soil	Depth (cm)	Sand/loam/clay (%)	Pore volume (%)	Bulk density (g/cm³)	pH	Field capacity (%)	Organic carbon content (%)
Sand							
I	0–25	94/6	43	1.5	6.5	14.8	0.667
II	25–75	93/7	35.7	1.7	5.8	9.5	0.21
III	75–200	96/4	39.6	1.6	6.3	5.0	0.0
Loam							
I	0–25	3/71/26	51	1.3	7.7	34.8	1.0
II	25–50	2/52/46	43	1.5	8.0	39.8	0.6
III	50–200	10/61/29	39	1.65	8.5	35.0	0.0

Table 2. Agricultural practices and climate data for the two scenarios

Soil	Crop	Application date	Harvest date	Period (days)	Yield (dt/ha)	Precipitation (mm)	Mean temp. (°C)
Sand	Winter barley	Apr. 10	Jul. 11	92	37	83	15
Loam	Winter wheat	Apr. 10	Aug. 5	117	94	178	15

An iterative procedure was applied to calculate the development of the biomass, the water content, and the volumes. After every time step (dt = 1 day), these properties are calculated and transferred to the plant model. Corn started to grow on the 44th day after pesticide spraying (May 24). Until this time, only stem and leaves were developing, assuming a biomass ratio of 1:1. After the 44th day only the corn grew. The curves were fitted to the different yields of the two crops. Figure 2 shows the development of the dry masses of winter barley and of winter wheat.

The plant model uses chemical concentrations on a volume basis, i.e., kg/m^3. The volumes V_i (m^3/ha) are calculated from the dry masses m_i (kg/ha) the water contents W_i (–), and the densities ρ_i (kg/m^3) (the index i indicates stem, leaves, and corn):

$$V_i = [m_i / (1 - W_i)] / \rho_i \tag{8}$$

D. SUBSTANCE PROPERTIES

Three pesticides were selected for simulation: carbofuran, isoproturon, and terbuthylazine. Carbofuran is a carbamate insecticide, isoproturon is a phenylurea, and terbuthylazine is a s-triazine herbicide. Common soil surface application rates of 1 kg/ha for isoproturon and terbuthylazine and 4 kg/ha for carbofuran are assumed. All three pesticides are water-soluble and moderately degradable organic chemicals. The physical and chemical substance properties are given in Table 4. Volatilization from soil is rather unlikely because of the low air/water partition coefficients (K_{aw}). Carbofuran and isoproturon have similar n-octanol/water partition coefficients (log K_{ow}), which indicates a similar uptake behavior into crops.

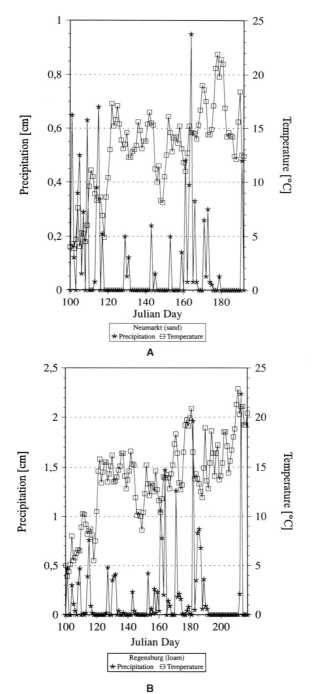

FIGURE 1. Precipitation and temperature during the simulation period: (A) Neumarkt (sand); (B) Regensburg (loam).

Table 3. Crop properties

Crop	Organ	Density (kg/m3)	Lipid content (%)	Water content (%)	Initial dry biomass (kg/ha)
Barley	Stem	1000	3	15–75	384
	Leaves	500	3	15–75	384
	Corn	1000	—	15	0
Wheat	Stem	1000	3	15–75	412
	Leaves	500	3	15–75	412
	Corn	1000	—	16.5	0

Pesticides were mainly chosen because their soil sorption and degradation are much better studied than for other organic chemicals. However, even for well-investigated pesticides these key properties often provide the greatest uncertainties in simulation models.[7] Since depth-dependent sorption and degradation were not measured, literature data had to be used. Only for terbuthylazine were linear equilibrium sorption coefficients (K_d values) measured for the sandy Ap horizon. Mostly K_{OC} values from batch studies can be found in the literature; these values have to be transformed into K_d values by multiplication with the organic carbon content of the various soil horizons. The extrapolation of batch K_d and K_{OC} to the field often results in overestimation of the adsorption.[8] Simultaneously, the fraction available for leaching and uptake is underestimated. The "best estimated" K_{OC} and K_d values are given in Table 4. Sorption is neglected in the third horizon because the organic carbon content is zero.

The degradation half-lives at 20°C were taken from the literature, preference being given to those values which were determined with comparable soils. Half-lives were doubled for the second horizon and set to zero for the third horizon (no humus content). The actual half-lives were adjusted to the soil temperature by using van't Hoff's rule (doubling of half-life with each 10°C reduction). Hydrolysis can be neglected within the more or less neutral pH range. Isoproturon is degraded much faster than the other two substances. Terbuthylazine sorbs stronger and is degraded less rapidly.

Concentration factors required for the calculation of translocation and fate of organics in stem, leaves and fruits are estimated from the log K_{OW}.[5,9] The transpiration stream concentration factor TSCF determines the uptake from soil water into the xylem. Carbofuran and isoproturon have TSCF values near the optimum (Table 4). The TSCF of terbuthylazine is lower because of the higher log K_{OW}. Organic chemicals can be metabolized by plant enzymes; half-lives given in Table 4 are estimates from literature data.

E. DATA TRANSFER FROM SOIL TO PLANT MODEL

As described above, the soil model calculates water and chemical uptake into the transpiration stream. The general data flow in the combined simulation model network SNAPS is shown in Figure 3. The input data file for the soil model contains the basic soil properties and the daily meteorological data; that

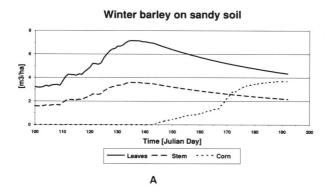

A

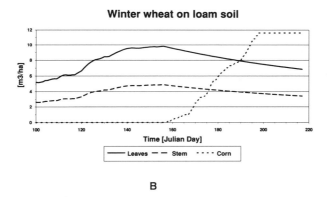

B

FIGURE 2. Development of dry biomass: (A) winter barley; (B) winter wheat.

for the plant model contains the crop properties, including crop growth. The substance data file provides the sorption coefficients, the partition coefficients K_{aw} and log K_{OW}, and the degradation half-lives in soil and crop. Moreover, after each time step several data are transferred from the soil model to the plant model. The data transfer can be "on-line" (i.e., by piping) or "off-line" (i.e., by writing a data file with subsequent file transfer). The following six data are transferred after each time step:

day: Julian day
Q_w: water uptake rate (m³/ha/s)
U_p: substance uptake rate (kg/ha/s)
T_a: air temperature (°C)
rh: relative air humidity (%)
LAI: leaf area index (−)

Total dry masses m_t are calculated from Q_w by adjusting the transpiration coefficient TC:

Table 4. Substance properties and application rates of the three pesticides

Property	Carbofuran	Isoproturon	Terbuthylazine
Sum formula	$C_{12}H_{15}NO_3$	$C_{12}H_{18}N_{20}$	$C_9H_{14}ClN_5$
Molar mass (g/mol)	221.26	206.29	229.5
Water solubility (mg/l)	700	72	8.5
Vapor pressure (Pa)	$2.7 \cdot 10^{-3}$	$3.3 \cdot 10^{-6}$	$1.5 \cdot 10^{-4}$
K_{AW} (–)	$3.5 \cdot 10^{-7}$	$3.9 \cdot 10^{-9}$	$1.67 \cdot 10^{-6}$
log K_{OW} (–)	1.82	2.3	3.06
K_{OC} (cm³/g)			
Sand	105	92	479
Loam	105	92	259
Kd (cm³/g)			
Sand I[a]	0.7	0.31	3.2
Sand II	0.24	0.19	1.1
Loam I	1.05	0.92	2.59
Loam II	0.68	0.55	1.55
$T_{1/2}$(20°C; days)			
Sand I	95.5	13	85
Sand II	191	26	170
Loam I	42.4	14	100
Loam II	84.8	28	200
TSCF (–) barley/wheat	0.784	0.686	0.409
$T_{1/2}$ (days) barley/wheat	4	7	7
Application rate (kg/ha)	4	1	1

Note: See Symbols section for definitions of properties.

[a] Soil horizon from Table 1.

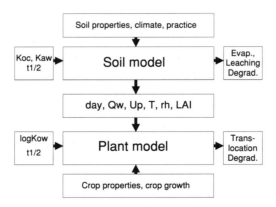

FIGURE 3. Data flow in the combined soil/plant model. Evap. = evaporation, Degrad. = degradation; see Symbols section for other definitions.

$$dm_t/dt = TC \cdot Q_w$$

with $m_t = 0.5 \cdot m_{stem} + 0.5 \cdot m_{leaves}$ until corn development (44th day) and with $dm_t/dt = dm_{corn}/dt$ until harvest.

The output file of the soil model contains the water dynamics, root development and chemical dynamics, and mass balance in soil. A separate output file is produced by the plant model and gives the results on the concentrations in stem, leaves, and corn. Mass balances in the plant are also calculated.

IV. RESULTS AND DISCUSSION

Chemicals are taken up with the transpiration water. The dynamics of soil and transpiration water, root development, and chemical transport determine the uptake and translocation of organic substances.

A. WATER BALANCE AND WATER UPTAKE

Both scenarios show a negative total water balance at harvest (end of the simulation period; Table 5). The water deficit is 127 mm for the first scenario (sand/winter barley) and 276 mm for the second scenario (loam/winter wheat), i.e., about 200% of the infiltrated water. Hence, a quantity equal to that of the infiltrated water is lost from the initial soil water content. The dominant loss process is the transpiration. Leaching and evaporation (i.e., water flow to the bottom and water loss from the soil surface) are only of minor importance.

The depth- and time-dependent water uptake into barley and wheat is shown in Figure 4. The pattern of water uptake is determined by the precipitation pattern (Figure 1) and the root development on sandy and loamy soil. Physiological differences of barley and wheat were neglected. Barley roots grew mainly in the upper layer of the sandy soil and took the main part of the water out of the upper soil layer (0 to 50 cm). Only after drying of the upper soil layer in the final 20 days before harvest is water taken up by roots to a 80 cm depth. The main water uptake occurred around the 30th day of the vegetation period. A second peak was observed around the 70th day. A drought between these two peaks caused a massive decrease of water uptake.

Wheat on loamy soil shows a more continuous pattern of water uptake over the soil profile and over time than barley (Figure 4B). The main reason is the higher field capacity of loam, which provides a better water supply. The main fraction of water was taken up at the end of the vegetation period. The difference in water uptake patterns is one of the reasons for the differences in substance uptake. The other reasons are the chemical transport in the soil column and the degradation of chemical during percolation.

B. CHEMICAL TRANSPORT, DEGRADATION, AND UPTAKE

Pesticides and other organic chemicals are transported in soil by a combined convective and dispersive/diffusive mass flow. They can be degraded by

Table 5. Total water balance for both scenarios

	Winter barley on sand	Winter wheat on loam
Precipitation (mm)	83	178
Interception (mm)	19	38
Infiltration (mm)	62	140
Transpiration (mm)	106	188
Evaporation (mm)	7	69
Leaching (mm)	14	19
Total deficit	127	276

microorganisms as well as by chemical hydrolysis. The scenarios assume a single application onto the soil surface with subsequent vertical migration under a specific climate regime. Figure 5 shows the time/depth profiles of the three pesticides. Because of the high water deficit in both soils and because of the effective microbial degradation, chemicals were not transported below the root zone; i.e., during the main vegetation period pesticides were not leached. This finding is valid for all three simulated pesticides in both soils.

The chemical mass balance at various times during the vegetation period is shown in Figure 6. The percentages are related to the total amount of the initial application. They indicate the variability of plant uptake and degradation in sandy and loamy soils. In most cases, uptake is below 10% at harvest.

Figure 7 shows the time- and depth-dependent uptake rates of the three pesticides into barley and wheat. The uptake rates are expressed in micrograms of chemical per square centimeter of soil per day. All three pesticides were mainly taken up within the first 30 days after spraying. A second maximum occured around the 60th day of the simulation period. The temporal uptake pattern is quite similar for all pesticides. It is determined by the water uptake on sand and loam. However, the heights of the uptake rates are different. Degradation results in lower uptake rates due to decreased concentrations in soil water. The simultaneous transport into deeper soil has no pronounced effect because transpiration is the dominant process of soil water loss.

Carbofuran has the highest uptake rates, whereas isoproturon and terbuthylazine show lower values. Isoproturon is rather water-soluble and sorbs only slightly, but is rapidly degraded. Uptake is therefore not as effective as for carbofuran, which has a higher degradation half-life in soil. Terbuthylazine is adsorbed and has the lowest TSCF value. Furthermore, degradation is lower than for isoproturon and carbofuran. Uptake is observed mainly from the upper soil layer, but with no decrease in the second part of the simulation period. The comparison of the three pesticides demonstrates the sensitive dependency of chemical uptake on the various dynamic processes.

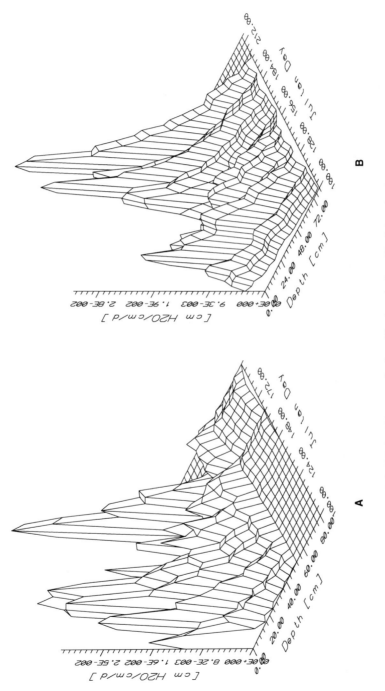

FIGURE 4. Water uptake into (A) winter barley; (B) winter wheat.

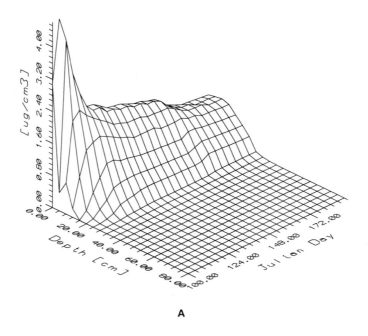

A

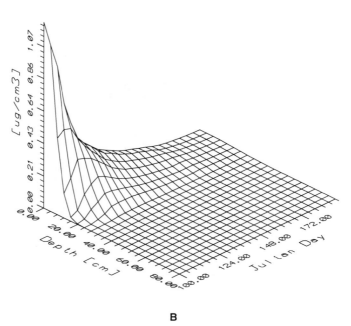

B

FIGURE 5. Concentration profile in soil: (A) carbofuran in sand, (B) isoproturon in sand, (C) terbuthylazine in sand, (D) carbofuran in loam, (E) isoproturon in loam, (F) terbuthylazine in loam.

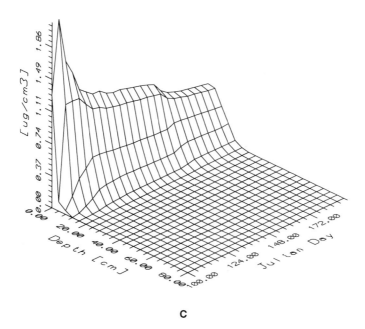

C

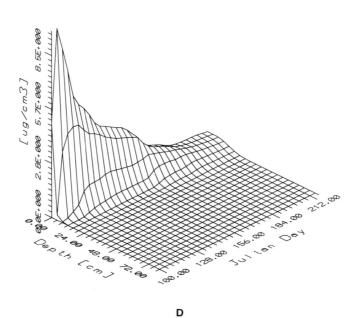

D

FIGURE 5 (continued).

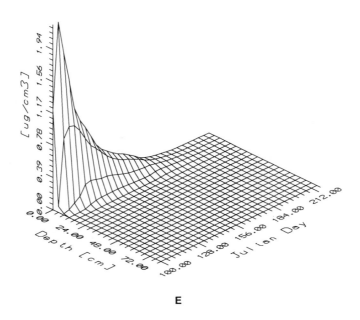

E

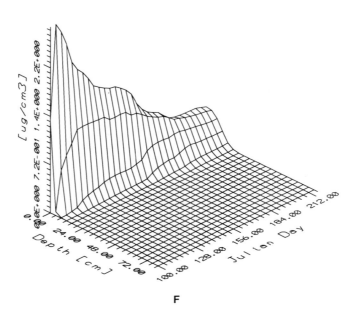

F

FIGURE 5 (continued).

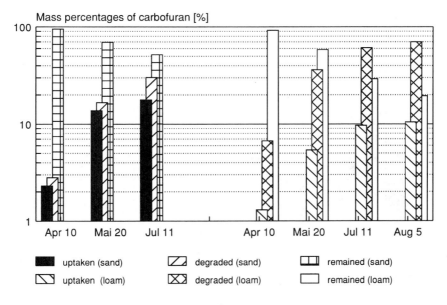

A

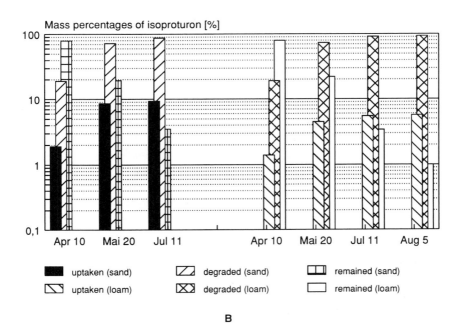

B

FIGURE 6. Chemical mass balance (percentages of chemical in soil, in plant, and degraded after various times): (A) carbofuran, (B) isoproturon, (C) terbuthylazine.

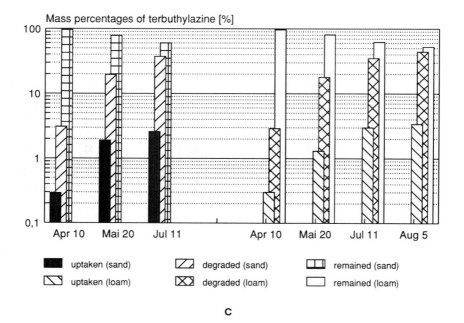

C

FIGURE 6 (continued).

C. TRANSLOCATION AND DISTRIBUTION IN PLANTS

After entering the xylem, chemicals are translocated from the roots and distributed to stem, leaves, and fruits (corn). The temporal patterns of concentrations on a fresh weight basis are given in Figure 8.

Leaves and stem are loaded with chemicals within the first half of the simulation period. After the 44th day (beginning of corn development) the concentration in corn increases and levels in stem and leaves decrease. The highest concentrations are observed for carbofuran in winter barley on sandy soil (maximum values above 10 mg/kg fresh weight in leaves). These high concentrations of the insecticide stem from an effective uptake (optimum log K_{ow}) and the high application rate (4 kg a.i./ha). Although isoproturon was applied at a lower rate (1 kg a.i./ha), its concentration in plants is only 2.5 times lower than that of carbofuran. The lowest concentrations are calculated for terbuthylazine (0.4 to 1.0 mg/kg fresh weight for leaves), which is slightly uptaken, but volatilized from foliage via the stomatal pathway to air.

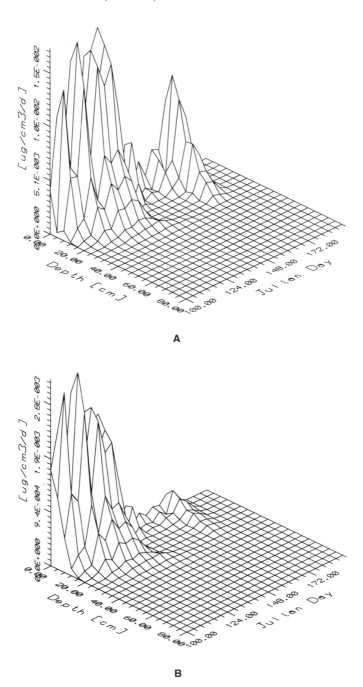

FIGURE 7. Chemical uptake rates: (A) carbofuran into barley, (B) isoproturon into barley, (C) terbuthylazine into barley, (D) carbofuran into wheat, (E) isoproturon into wheat, (F) terbuthylazine into wheat (units: μg/cm³ soil/day).

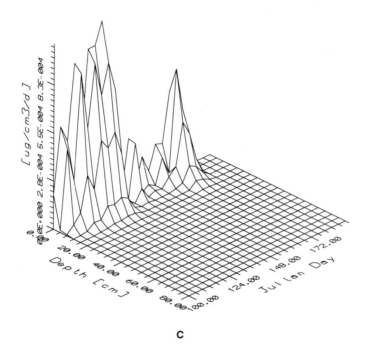

C

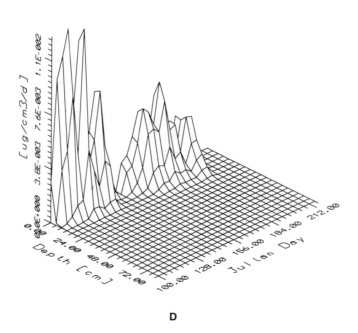

D

FIGURE 7 (continued).

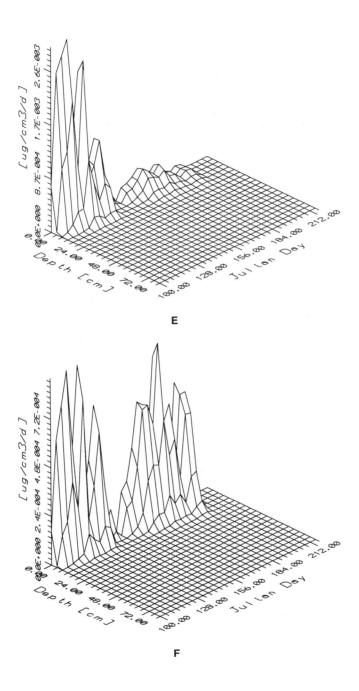

FIGURE 7 (continued).

In Figure 9 the mass balances at harvest for the barley/sand and wheat/loam scenarios are shown. Remarkable quantities are volatilized for all chemicals (between 5 and 43%). Trapp[6] discusses the three resistances which determine the leaf-air exchange: stomatal, cuticular, and atmospheric resistance. Their contributions depend on the substance-specific partition coefficients, in particular on the ratio of K_{OW} to K_{aw}, and on the plant-specific conductivities. The quantities remaining in stem, leaves, and fruits are rather low. They lie in general below 1% of the total amount taken up. Degradation in plant is the dominant elimination process, accounting for between 57 and 95%.

The contamination at harvest, in particular in corn, is most interesting for the assessment of risks posed by food and feed supply. Due to the degradation within the plants and the volatilization from foliage the concentrations in corn decrease until harvest. However, at the time of harvest pesticide concentrations in the range of about 0.1 mg/kg were calculated in corn. The maximum allowed concentration in cereals according to the regulations in Germany is 0.2 mg/kg for isoproturon and 0.1 mg/kg for terbuthylazine.[10] For carbofuran the maximum allowed concentration in maize is 0.2 mg/kg.[10] Calculated residue concentrations are of the same order of magnitude as the maximum allowed concentrations. The standard for pesticide concentrations in drinking water in Germany is much lower (1 ppb for a single pesticide and 5 ppb for the sum) and is independent of the toxic properties of the particular pesticide.

V. CONCLUSIONS

The systemic uptake and translocation of pesticides under the changing climate and water regime and under simultaneous pesticide leaching and transformation during the vegetation period can be adequately described with the simulation models incorporated in the SNAPS model system. The effectivity of uptake from soils with different field capacities was shown for sandy and loamy soils. Time series of the concentrations in the various parts of soil and plant can be easily calculated. They can be used to evaluate desired levels for insecticide or herbicide action or to assess undesired residues of toxicants, e.g., in food and feedstuff. Rates of losses by degradation and volatilization give information on the time course of removal. The simulated scenarios demonstrate that pesticides with a groundwater contamination potential[11,12] also have a remarkable potential for crop contamination. Although the concentrations at harvest are lower than the maximum allowed concentrations, the peak concentrations in the course of the vegetation period are significantly higher (by a factor of 50 to 100). This was demonstrated especially for carbofuran and isoproturon.

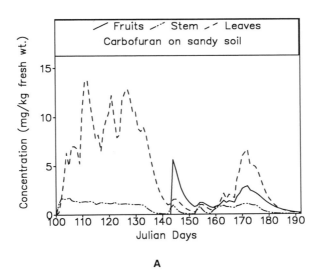

A

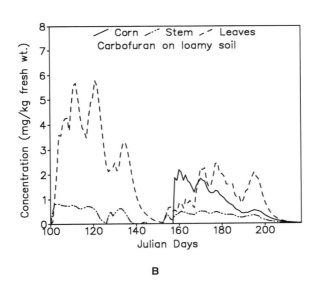

B

FIGURE 8. Temporal pattern of concentrations in stem, leaves, and corn: (A) carbofuran in barley, (B) isoproturon in barley, (C) terbuthylazine in barley, (D) carbofuran in wheat, (E) isoproturon in wheat, (F) terbuthylazine in wheat.

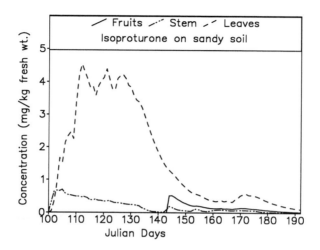

C

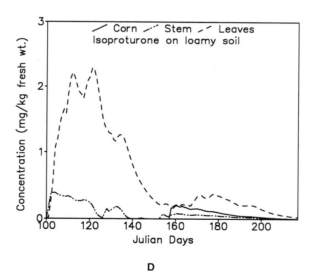

D

FIGURE 8 (continued).

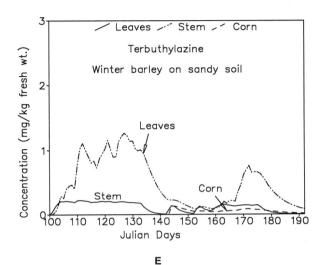

E

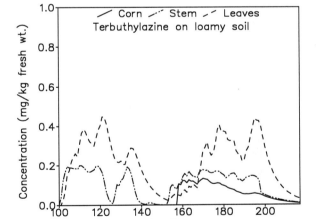

F

FIGURE 8 (continued).

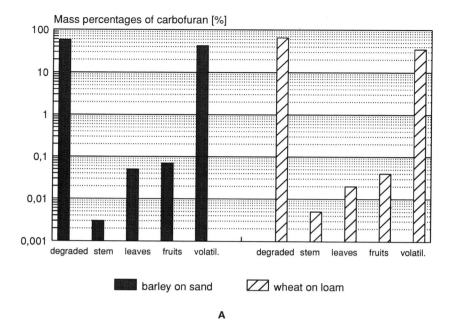

A

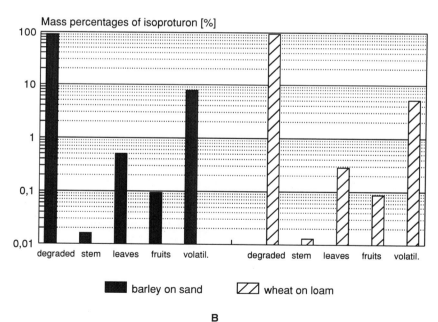

B

FIGURE 9. Chemical mass balance in crop at harvest (percentages of chemical in stem, leaves, corn; amount degraded and volatilized): (A) carbofuran, (B) isoproturon, (C) terbuthylazine.

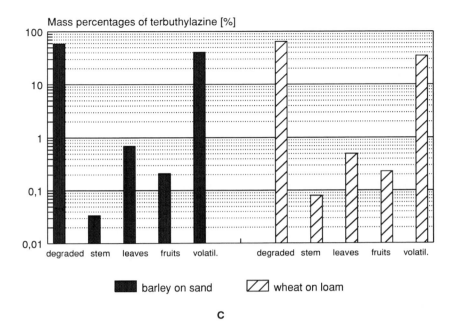

FIGURE 9 (continued).

Analysis and simulation are not restricted to pesticides. Organic chemicals with similar properties will behave in a similar manner. The pathways of transport, leaching, uptake, and volatilization are determined mainly by the substance-specific partition coefficients. The quantities and, thus, the concentrations depend on the various processes of soil water dynamics, plant development and properties, sorption, and degradation. For more lipophilic chemicals other pathways e.g., volatilization from soil surface or direct uptake from the atmosphere can become much more important than for the water-soluble pesticides studied here.[13]

ACKNOWLEDGMENT

The authors would like to thank Dr. Stefan Trapp for the calculation of translocation and distribution in cereals and Markus Morgenstern and Hubert Steindl for providing soil profile data.

SYMBOLS

Symbol	Definition	Units
$f(z)$	Empirical depth-in-soil function	–
$g(T_s)$	Temperature correction by van't Hoff's rule	–
m_i	Dry mass of plant compartment; index i indicates stem, leaves, and corn	kg/ha
m_t	Total dry mass of plant	kg/ha
q	Darcy water velocity in soil	cm/day
rh	Relative air humidity	%
t	Time	days
$w(z,t)$	Root length distribution	cm/cm^3
z	Depth in soil	cm
z_{root}	Rooting depth	cm
C	Dissolved concentration	$\mu g/cm^3$
C_{apw}	Differential water capacity	$1/cm\ H_2O$
C_T	Total chemical concentration in soil	$\mu g/cm^3$
$D(\theta,q)$	Effective diffusion dispersion coefficient	cm^2/day
K	Soil hydraulic conductivity	cm/day
K_{aw}	Air-water partition coefficient	–
K_d	Linear equilibrium adsorption coefficient	cm^3/g
K_{oc}	Organic carbon-water partition coefficient	cm^3/g
K_{ow}	n-octanol-water partition coefficient	–
LAI	Leaf area index	–
OC	Organic carbon content in soil	–
Q_w	Plant water uptake rate	$m^3/ha/s$
S	Root water uptake rate	cm H_2O/cm/day
S_c	Chemical sink term (degradation, root uptake)	$\mu g/cm^3/day$
S_{deg}	Chemical sink term for degradation	$\mu g/cm^3/day$
S_{root}	Chemical sink term for root uptake	$\mu g/cm3/day$
T_a	Temperature in air	°C
T_s	Temperature in soil	°C
TC	Transpiration coefficient	kg/m^3 H_2O
$T_p(t)$	Potential transpiration	cm H_2O/day
TSCF	Transpiration stream concentration factor	–
$T_{1/2}$	Chemical half-life in soil or plant	days
U_p	Substance uptake rate	kg/ha/s
V_i	Volume of plant compartment; index indicates stem, leaves, and corn	m^3/ha
W_i	Water content on mass basis of plant compartment; index i indicates stem, leaves, and corn	–
$\alpha[\Psi(z,t)]$	Empirical reduction function in case of water stress	–
μ_0	Bulk first-order degradation rate at 20°C in upper soil horizon	1/d
ρ	Soil bulk density	g/cm^3
ρ_i	Density of plant compartment	kg/m^3
θ	Volumetric water content	–
Ψ	Soil matrix potential	cm H_2O

REFERENCES

1. Feddes, R.A., P.J. Kowalik, and H. Zaradny (1978), *Simulation of Field Water Use and Crop Yield,* PUDOC, Wageningen, The Netherlands.
2. Wessolek, G. and S. Gäth (1989), Integration der Wurzellängendichte in Wasserhaushalts- und Kaliumanlieferungsmodellen, *Kali-Briefe,* 19: 491.
3. Behrendt, H., M. Matthies, H. Gildemeister, and G. Görlitz (1990), Leaching and transformation of glufosinate-ammonium and its main metabolite in a layered soil column, *Environ. Toxicol. Chem.,* 9: 541.
4. Campbell, G.S. (1985), *Soil Physics with Basic,* Elsevier, New York.
5. Briggs, G.G., R.H. Bromilow, and A.A. Evans (1982), Relationships between lipophilicity and root uptake and translocation of nonionised chemicals by barley, *Pestic. Sci.,* 13: 495.
6. Trapp, S. (1995), Model for uptake of xenobiotics into plants, in *Plant Contamination: Modeling and Simulation of Organic Chemicals,* J.C. Mc Farlane and S. Trapp, Eds., Lewis Publishers, Boca Raton, FL, Chap. 5.
7. Villeneuve, J.P., P. Lafrance, O. Baton, and P. Frechette (1988), A sensitivity analysis of adsorption and degradation parameters in the modeling of pesticide transport in soils, *J. Contam. Hydrol.,* 3: 77.
8. Schernewski, G., N. Litz, and M. Matthies (1990), Untersuchungen zur Anwendbarkeit von Sorptions koeffizienten für die Simulation der Verlagerung von 2,4,5-T und LAS in Böden, *Z. Pflanzenernaehr. Bodenkd.,* 153: 141.
9. Trapp, S., M. Matthies, I. Scheunert, and E.M. Topp (1990), Modeling the bioconcentration of organic chemicals in plants, *Environ. Sci. Technol.,* 24: 1246.
10. Perkow, W. (1988), *Wirksubstanzen der Pflanzenschutz- und Schädlingsbekämpfungsmittel,* Verlag Paul Parey, Berlin.
11. Matthies, M. (1987), Fate modeling of pesticides in groundwater, in *Pesticide Science and Technology,* R. Greenhalgh and T.R. Roberts, Eds., Blackwell Scientific, Oxford, pp. 373–380.
12. Klein, A.W., P. Apel, and J. Goedicke (1992), UBA — principles on criteria and procedures for environmental assessment of pesticides, *Chemosphere,* 24: 793.
13. Trapp, S., M. Matthies, and A. Kaune (1994), Transfer von PCDD/F und anderen organischen UmweltchemiKalien in System Boden-Pflanze-Luft, I. Modellierung des Transferverhaltens, *UWSF-Z. Umweltchem. Ökotox.* 6(1) 31–40.

Index

Index